ECLIPSE

ECLIPSE

Journeys to the Dark Side of the Moon

FRANK CLOSE

OXFORD
UNIVERSITY PRESS

OXFORD
UNIVERSITY PRESS

Great Clarendon Street, Oxford, OX2 6DP,
United Kingdom

Oxford University Press is a department of the University of Oxford.
It furthers the University's objective of excellence in research, scholarship,
and education by publishing worldwide. Oxford is a registered trade mark of
Oxford University Press in the UK and in certain other countries

First Edition published in 2017

Impression: 1

Published in the United States of America by Oxford University Press
198 Madison Avenue, New York, NY 10016, United States of America

British Library Cataloguing in Publication Data
Data available

Library of Congress Control Number: 2016946808

ISBN 978–0–19–879549–0

Printed in Great Britain by
Clays Litho Ltd, St Ives plc

To my Teachers

'Everything under the sun is in tune
But the sun is eclipsed by the moon.'
Pink Floyd

Photo of total eclipse (photo: Frank Close). See Plate 1 for a colour version.

PREFACE: TOUCHING INFINITY

My teacher held a sycamore seed in her fingers, twisted it, and let go. I watched, fascinated as it floated gently to the earth, its two delicate miniature wings rotating like the blades of a helicopter. It was my first day at infant school. She promised that if I planted the seed in a pot of earth, a tree would emerge after seven years.

I did. It didn't. Seeds need food and water no less than do small boys, and the importance of these crucial ingredients was overlooked. And at four years old, it is hard to imagine seven years into the future.

That was my first memorable introduction to science. The second was more successful.

By this stage I was eight years old. My teacher that year—Mr Cyril Laxton—was no scientist, however. He sang alto in a choir, and was one of three brothers, each of whom was a gifted sportsman. One day he used a gleaming red cricket ball and a leather football to introduce me to one of nature's most remarkable phenomena: a total eclipse of the sun. He promised that if I waited 45 years, I could see the real thing for myself, and it would be wonderful.

At eight years old, to imagine 45 years in the future is to touch infinity.

But I waited. And he was right. His intervention defined my life.

This is the story of how.

ACKNOWLEDGEMENTS

Eclipse is about my lifelong fascination with eclipses, and my journey to solve a puzzle—how did the sun stop and the moon turn back during a solar eclipse, as recorded in the Book of Joshua? I am indebted to Graham Farmelo for having suggested one sunny day in the centre of the universe that I write this personal history. It could not have happened but for the inspiration of Cyril Laxton, my teacher in primary school—and to avoid a spoiler alert, you will have to read to the very end to understand his complete role in this tale. Among many who have shared their experiences of eclipses in far flung parts of the globe, have read the manuscript in part or in whole, or taught me about eclipses, my thanks are given to

Marilyn and Peter Boldon
Sheila Cavanagh
John Charap
Michael Duff
Bill and Cathy Colglazier
Michael Collins
Alex Fillipenko
Tim Hurst Brown
Jeri Johnson
Bill Kramer
Rolf Landua

Stephen Leonard
John Mason
Alison Miller
Landon Curt Noll
Hans Schmidt
Emma Smith

Thanks to my agent Patrick Walsh and editor Sonke Adlung, for bringing Dark Side to the light of day, and to my family, for having experienced my enthusiasm and sometimes shared in it, and to innumerable eclipse chasers with whom I have spent time in the dark. And finally, if this history should somehow magically reach an anonymous young man from Zambia, who experienced totality as a boy in 2001, then I hope that you indeed: 'believe in science'.

CONTENTS

Prologue: Dark Amid the Blaze of Noon

What is the most beautiful natural phenomenon that you have ever seen? A brilliant rainbow set against a distant storm, the shimmering aurora in an Arctic night, or a blood red sky just after sunset, perhaps? Anyone who has experienced the diamond ring effect that heralds the start of a total solar eclipse, however, will tell you that it puts all others in the shade.

That this marvel happens is thanks to a cosmic coincidence: the sun is both 400 times broader than the moon and 400 times further away. This makes the sun and moon appear to be the same size. So if the moon is in direct line of sight of the sun, it can completely and precisely block it from view.

A total eclipse of the sun happens about once every 18 months. As the moon moves slowly across the face of the sun, it casts a shadow on the earth's surface (see figure), about 100 miles in diameter. As our planet spins in its daily round, the moon's silhouette rushes across land and sea at about 2000 miles an hour.

Total eclipse over the Sahara. A meteorological satellite's view of the moon's shadow as it crossed the Sahara Desert during the solar eclipse of 29 March 2006. Had this been a high-resolution image from a spy satellite, the author would be visible near the centre of the silhouette. The full story is in Chapter 7. A moving image of the shadow is at http://bruxy.regnet.cz/web/meteosat/EN/solar-eclipse-2006/ (© 2006 EUMETSAT, Meteosat 8 image received and processed by Martin Bruchanov.) See Plate 2 for a colour version.

There is a slow build-up to the totality show, as the moon gradually covers the sun, which becomes a thin crescent as twilight falls. As the climax approaches, excitement mounts. The temperature drops, and then, in the west, a wall of darkness like a gathering storm rushes towards you. This apparition is the moon's shadow. In an instant you are enveloped by the gloom. The last sliver of sun disappears and, as from nowhere, a diamond ring flashes around a black hole in the sky, vibrant, like a living thing.

For those beneath the shadow as it passes, the sounds of animals cease, and life seems in suspended animation as for a few minutes night comes to the dome of the sky directly overhead, and covers the land from one horizon to the other. Look up myopically, and you would see stars as if it were normal night, accompanied by an awesome sight: that inky circle, surrounded by shimmering white light, like a black sunflower with the most delicate of silver petals. One watcher has described it to me as like 'looking into the valley of death with the lights of heaven far away calling for me to enter'.

After the thrill of an eclipse you can't wait to do it again, but wait you must until that exquisite alignment of sun, moon, and earth comes around once more. When it does, you must go to the thin arc where the moon's shadow momentarily sweeps across a small part of the globe. For a total eclipse is only visible at special places on earth; a mere 0.5% of the earth's surface is totally obscured by the moon's shadow for just a few minutes, while the remaining 99.5% sees either a partial eclipse or nothing at all. Stay at home and you will miss it.

On 21 August 2017, up to 200 million people will gather in a narrow belt across the USA, from Oregon to South Carolina, to witness the most watched total solar eclipse in history. Anyone who hasn't experienced totality might struggle to understand why people are prepared to adventure to the far side of the earth, by plane, ship, even on the hump of a camel, to be there.

I first learned about total eclipses six decades ago. I never expected that half a century would pass before I myself saw one. Far less did I anticipate that I would then spend the latter years of my life planning expeditions throughout the globe to watch them.

I

Peterborough 30 June 1954

When I was a small child, I was told a story in which, 'once upon a time', a piece of the sky became detached in the middle of the day. The errant disc, no bigger than the size of a fingernail at arm's length, was so small that at first no one noticed. But then it began to slide over the face of the sun. Initial wonder soon turned to panic as what was at first a mere blemish in the outer rim continued to devour the golden orb. The once bright ring gradually reduced to a thin crescent. Finally even that sliver evaporated as the disc covered the sun like a lid, and the earth was plunged into darkness.

The first clue that this story might be fact came in midsummer, 1954, when I was eight years old.

My childhood home, Peterborough, was a two-dimensional town on the flatlands at the western edge of the Fens. Located immediately north of the belt of clay from which the London Brick Company produced blocks the colour of the setting sun, Peterborough consisted of thousands of red houses. Five mornings a week I walked the mile through streets of cloned semi-detached redness to St Mark's County Primary School for Boys, in Gladstone Street. And on five afternoons I would saunter back home again, occasionally pausing around the corner from the school to watch a blacksmith smelt horseshoes in a white-hot forge, as bright as the sun itself.

St Mark's dated from the Victorian era, its buildings a single storey version of Blake's dark satanic mills. It was near the main line of the London and North-Eastern Railway, where the whistles and smoke from the likes of the *Flying Scotsman*, or of the world's fastest steam engine, *Mallard*, were regular features of the school day.

Boys from age 7 to 11 were grouped into classes of 50. The classroom had a high ceiling, with small windows near the roofline, so far above our heads that even on tiptoe all we could see was the sky. The room had no direct access to the outside world, and smelt dusty, like a crypt, bereft of fresh air. It was so gloomy that three 100 watt tungsten light bulbs, which hung from the ceiling, were lit even on sunny days. The environment felt like prison, where we were cut off from the external world.

The linoleum floor was covered with rows of wooden desks, whose roughened lids would open to reveal our exercise books, and pens whose nibs were invariably crossed. Atop each desk was an inkwell, which would be filled with blue powdery liquid each morning by the ink-monitor, the best pupil of the previous week. On the front wall there was a huge board. Our teacher, Mr Laxton—a lean wiry sporty man with pince-nez spectacles,—would cover this with chalk as we were drilled with facts for an hour and a half: historical events, multiplication tables, and how to write neatly on a straight line.

Lessons began at nine o'clock, and continued until about 10.30 when a bell would ring to announce a 30 minutes mid-morning break. After our half-an-hour of liberty in the fresh air, periodically polluted by the sensuous smell of smoke from the railway, the bell would sound once more, and lessons would re-commence. One of the few historical dates that I could remember was that of Magna Carta, as 12.15 was when our lunch-break began. The ubiquitous bell would ring again, audible throughout the school, to announce our temporary freedom. This bell was a permanent fixture, as regular as the chimes of Big Ben, which together with the sound of pips on BBC Radio formed the metronome for British life in the 1950s. It would ring again at a quarter past one to announce the start of classes for the afternoon.

Except that on Wednesday 30 June, 1954, the normal order didn't happen.

This had begun as a standard day, one where in the morning I had trudged reluctantly to school, as usual. By the evening, however, my world had changed as I returned home excited. The events of that day had ignited what would become my lifelong passion for science.

* * * * *

It was just nine days after Midsummer, and the sun had reached its zenith. There were a few white fluffy clouds in the sky, but the air was still, the sun shone bright, and it was warm: over 70° in the old Fahrenheit scale of temperature. I know, because I was given the task of recording it.

At the end of morning break that day, the bell remained silent. Instead we were told to line up in groups to hear an announcement from 'Pop Hayes', the Head Teacher. To boys who were not yet ten years old, the Head looked ancient, plump, with several chins and a thick moustache. In reality he was probably a sprightly fifty-something, as he addressed us like a sergeant-major giving orders to ranks of infantry. He barked that lunch would begin at noon, and that we were to reassemble in the playground at one o'clock to watch an eclipse of the sun.

At that moment I had no idea what this meant, except that possibly the afternoon would be spent outdoors, in the sunshine, rather than in a gloomy classroom. This release from purgatory was unique in my four years at the school, which is another reason why the day became so memorable.

And so, after a quick lunch, we lined up in the schoolyard once again. There were half a dozen rows of boys in short trousers (grey), socks (grey), shirts (grey), and sandals (scuffed). Our hair was cut uniformly: short back and sides. We must have looked like ranks of miniature soldiers facing south.

The eclipse was already under way. Pieces of darkened glass were handed out, through which we squinted gingerly at the crescent sun. Then we ignored our teachers' warnings, and took furtive glances at its brilliance with our naked eyes, daring ourselves to risk blindness. Having had the experience of seeing a partially eclipsed sun, everyone was ordered to return to their classrooms for lessons.

For a moment I thought my dream of an afternoon of freedom had gone. However, I was in luck. Mr Laxton became instantly popular by telling my classmates and me to remain outside.

He went into the school, briefly, and returned with a telescope. Having surrounded its shaft with a sheet of card, and mounted the tube with its collar onto a tripod, he angled the eyepiece so that it pointed at the sun. On the ground he placed a white sheet, which acted as a screen. The result was remarkable. The card cast a shadow on the sheet, and in the middle of this silhouette was a lustrous crescent: an image of the partially eclipsed sun. We could now watch the eclipse without needing to look at the sun directly, and the image on the ground was huge, larger even than the orb in the sky itself.

He explained that solids leave shadows, but that any small hole can cast an image, and so introduced us to the idea of the pinhole camera. First he demonstrated this with a sheet of paper, into which he had made a small hole. Although the hole was circular, like magic it revealed the image on the ground of a bright crescent sun. 'You can do it with your hands' he said, and showed us how. Placing his fingers as if in prayer, he then turned one hand sideways, so that it was perpendicular to the other. Then he very slightly spread his fingers, which opened small gaps between them. A handmade pinhole camera resulted (see also Figure 1A).

Most memorable of all were the trees. Dappled sunlight shone through their shady overhang. In the gaps between the shadows of the leaves were innumerable images of the sun.

When he had completed these demonstrations, we separated into small groups, with different tasks. One team was given sheets of white paper and made a record of the eclipse's progress, by tracing the telescopic image of the changing crescent at intervals during the next two hours. Another group made cut-out images of clock-faces, on which they positioned the hands to show the time

Figure 1A. Shadows and crescents. The lattice of a chair creates a matrix of pinhole cameras, which project crescent images of a partially eclipsed sun onto the floor (Photo: Frank Close). See Plate 3 for a colour version.

when each measurement was made. I was with another handful of boys who did 'real science': with a mercury thermometer, which was hung on the school wall, we recorded the temperature during the eclipse.

It was a warm, even hot, sunny day, or had been at noon. But instead of getting warmer as the sun reached its zenith, as on a normal day, the air cooled as the moon obscured the sun. By half past one in the afternoon, at the height of the eclipse when only a thin arc remained, the thermometer recorded a mere 50°Fahrenheit. Even boys used to washing in cold water—central heating was yet to reach the outposts of East Anglia—felt the chill. Then the sun began to grow and, after a few minutes, warmth began to return.

Mr Laxton explained what had happened. He said that the moon had crossed the line of sight to the sun, and hidden some of it: we had seen a 'partial eclipse' of the sun. Then, he added, if we had lived about 500 miles further north, in the Shetland Islands, the alignment would have been perfect and the sun totally obscured. A total eclipse of the sun, he informed us, was a sight to behold, and if ever we were lucky enough to see one, the experience would be unforgettable. He couldn't explain why, except to say that if it happened, we would know what he meant.

He promised to tell us more the next day. The first lesson on Friday would be about the stars.

* * * * *

The eclipse was a major event. That evening, the news on the radio reported that people in the north of Scotland had been plunged into darkness; the next morning there was even a photograph of an eclipsed sun in the newspaper.

When I arrived at school, I found pictures of the eclipse posted on the notice board. Mr Laxton had mounted a chart on the wall behind his desk at the front of our classroom. This display, which stretched along half of the entire wall, showed the results of our observations from the previous day.

From left to right, the images of the clock-faces showed the passage of time, from the start of the eclipse, to its midpoint, and onwards to the end. Beneath these he had pasted our sketches of the eclipse at the corresponding moments. Completing the display, and stuck above the clocks and crescents, was a long sheet of white paper on which was drawn a red line. The curve meandered down gently from the left hand side, and then rose up again gradually to the right, like the cross-section of a valley. He explained that this represented our measurements of the temperature throughout the eclipse.

There was excitement simply in knowing that we had been involved. Temperature, eclipse, and the time had been combined into a single story. These three independent endeavours were now joined, and revealed a great truth: the temperature went down and up as the visible sun first diminished and then grew back to a full circle.

He told us that this was scientific investigation. We were like detectives, who had gathered clues, and discovered a hidden message: the sun provides both heat and light. Cut out the sun and not only does the day turn gloomy, but it also becomes cold. When the sunshine returns, so does the warmth—but not at once. The air had continued to cool after the peak of the eclipse, and only begun to heat up again slowly. He said that as it takes a while to get warm after coming indoors in winter, so it takes a few minutes for the sun's heat to transfer to the air.

Nothing of what we did or revealed was the greatest of insights, perhaps, but it was true nonetheless. In a world of opinion, debate, uncertainty, this was fact, deduction, scientia—knowledge.

We had discovered an ability to understand, explain, and to predict. The sun rises each day, does it not; the moon goes through its phases every month; eclipses happen once in a while. Mr Laxton now helped us take the first steps along the eightfold way to enlightenment, and tried to explain how the eclipse had come about.

The world is round; it orbits the sun once every year and spins like a top making one revolution every 24 hours. Trapped by gravity on the surface of this vast merry-go-round, we are turned to face the sun by day, and whirled into the earth's shadow by night, over and over again. He illustrated this with a model. Being a sportsman, he chose a football to represent the earth, and a shiny new cricket ball for the moon. He completed the trio of essential players with a floodlight, which simulated the sun. Viewed from our perspective, the three were in a line. The floodlight was nearest to us, the football was next, and Mr Laxton, holding the cricket ball behind them, was back against the wall. He now pulled curtains over the windows to add to the room's natural gloom, and switched on the floodlight. The mock sun cast a shadow of the football onto the wall. Carefully holding the cricket ball—the moon—behind the football but above its shadow, he showed us how the front of the cricket ball was fully lit. This was 'full moon', he said.

'All light comes from the sun', he said. 'The moon doesn't shine, nor does a cricket ball, but they can both reflect light like a mirror.' This he now demonstrated to us. First, he moved the 'moon' around to the side of the 'earth', so that the floodlight illuminated the moon at a glancing angle. This, he explained, is what happens as the real moon orbits the earth: the amount of reflected light visible from the earth changes throughout the orbit, which makes what we call the 'phases of the moon'.

The angle of the moon relative to the sun thus determines how much of the lunar face is bright and how much dark. When the moon is on the opposite side of the earth to the sun, its face is full on to the sun and earthlings see the entirety illuminated. This is what we call full moon. About a fortnight later, from our perspective the moon is in the same area of the sky as the sun, and only a sliver reflects sunlight in our direction. This, he said, is known as 'new moon'. Then, with each subsequent day, the moon changes its location relative to the sun and reflects progressively more sunlight in our direction. He explained that it takes about 28 days for the moon to make a circuit, which is why there is a full moon every month and a new moon, midway between: a fortnight before and after.

He had been holding the cricket ball high in the air as he said this. The lamp was so bright that the shadow of the ball was visible on the wall. Now he lowered the ball, held by his fingertips, until it was in a direct line between the lamp and the football. The silhouette of the cricket ball now fell on the surface of the football—the shadow of the 'moon' was on the surface of the 'earth'. Any people on the real earth would thus see the sun blocked out when the shadow engulfed them. This, he explained, was an eclipse.

Next, he took the 'moon'—for that in our imagination is what the cricket ball had by now become— half an orbit to the rear and lowered it into the earth's shadow. Instead of a full moon, shining bright, he said, the moon would be obscured by the earth's shadow: a total lunar eclipse.

As I recall these events, decades later, I realise with hindsight that there were many questions that we should have asked, but didn't. This was all so new that I, at least, was entranced with the idea of balls that represented the moon and earth, and so brilliantly explained the phases of the moon: the eclipse suddenly appeared obvious. Or was it? One of my classmates saw what the rest of us had either overlooked, or been afraid to ask: why isn't there a total eclipse of the sun every month, and one of the moon regularly midway between them?

Mr Laxton said that the moon's orbit moves up and down, like when we were on a swing. Usually at new and at full moon, the moon passes above or below the direct line of sight, but once in a while it lines up, like had happened the day before. Now we were hooked. How long would it be before we could do this again?

At any random point on the earth's surface, a total solar eclipse will happen about once every 400 years on the average, but the bad news was that total solar eclipses, like good weather and sleek sports cars, were things that tended to be found abroad, away from England, in more exotic locations. The south seas, and places in the glorious British Empire, those assorted splodges of pink on the Mercator Projection on the wall of our classroom, seemed to have total eclipses on a regular cycle, whereas Peterborough had not been on an eclipse track since 1715. There's no point in sitting at home in the hope that one will come to you. Instead, you must go and meet one.

Of course, as in all games of chance, some are luckier than others. In 1724 another total eclipse had narrowly missed Peterborough, but residents of the south and west of England could have enjoyed them both. Years later I learned that people who lived near the Victoria Falls in central Africa around 2001 could have seen two total solar eclipses in as many years, and there was similar good fortune in Turkey in 1999 and 2006. But for most locations, several centuries can pass before a total eclipse occurs. Were we to sail the Pacific Ocean, we might happen across a total solar eclipse now and then, but to paraphrase Eliza Doolittle: in Hertford, Hereford, and Hampshire, eclipses, like hurricanes, hardly ever happen.

How can we tell when and where to go to have the experience? These questions began to interest me only years later. To answer them required understanding what defines a 'lunar month'.

Like an aged judge, who keeps asking counsel 'what exactly do you mean by X', where X is something that might have appeared to be blindingly obvious until the old sage raised the question, so the measure of the moon also turned out to be full of surprises. The definition of a lunar month, at first sight utterly trivial, turns out to be tricky to specify precisely. The problem is how to decide when the moon has returned to the 'same' position, to complete its circle of the earth, while the pair of them is also in orbit around the sun. To illustrate the conundrum, and also give the solution, let's transport the heavens onto a clock-face, with the sun at its centre. (See Figure 1B. I hasten to add: this is not to scale!)

For our purposes, the sun is at the centre of everything, and the orientations of the 12 remote zodiacal constellations are spread around the circumference. The position of the earth is at the tip of the hour hand, which rotates around the circle at the rate of one orbit per year. The 12 months correspond to the hourly numbers, from 1 for January to 12 for December. The orbit of the moon, meanwhile, makes small circles around the tip of the pointer—the earth.

In late December, the earth is atop the pointer at 12. Viewed from earth, the sun is aligned with the constellation Sagittarius, hence the terminology of astrologers: 'the sun is in Sagittarius'. The new moon is when the moon is at position ND—'New moon December'—in the illustration. If at this point the moon were

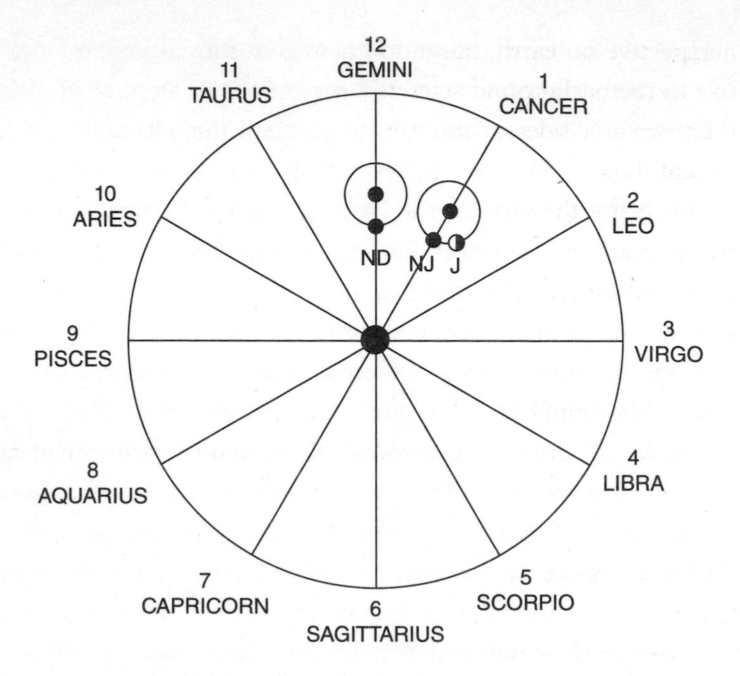

Figure 1B. Lunar months: synodic and sidereal. The sun is at the centre of the clock-face, with the 12 constellations of 'fixed stars' around the rim. The earth is the dark blob at the centre of the two circles below the 12 and the 1. The moon is the blob that orbits around the earth. The position ND denotes the new moon in December, when the moon is in direct line between earth and sun. In January, the earth had moved to the radial line centred on 1. The position J denotes the moon having reached the same point, relative to the fixed stars, as it was in December. However, new moon will not occur until it reaches the point NJ.

exactly in the plane of the clock-face, it would eclipse the sun. If the moon is slightly above or below this plane, there is no eclipse and we have the familiar new moon.

From the position ND, it takes about 27 and one-third days for the moon to arrive at position J on the clock-face. From our perspective on earth, the moon here is at the same point relative to the background stars as it was at ND in December. This is known as a 'sidereal' month—sidereal meaning 'relative to the distant stars'.

This is not the same as the time that elapses between successive new moons, however. During the intervening 27 days, the earth and moon have moved around the clock-face. As a result, our perspective on the distant stars has changed slightly; the sun is now 'in Capricorn'. The new moon does not occur until the moon reaches the point NJ.

This further distance corresponds to about one-thirteenth of its orbit—that is the ratio of 27.3 days of orbit around the clock-face, relative to the 365.2 days for a complete circuit. Put more directly: the time between successive new moons is about 2.3 days longer than a sidereal month. This is known as the 'synodic' month, from the Greek for 'meeting', and on the average lasts about 29 and one-half days.

The phases of the moon depend on the moon's orientation with respect to the sun as seen from the earth. From new moon, when the moon emerges from the line of sight of the sun, the moon waxes for two weeks, until, following full moon, it wanes, eventually forming a new 'new moon' after a synodic month.

A lunar eclipse is only possible at full moon, and it occurs if the trio of sun, moon and earth are so well aligned that the moon falls into the earth's shadow. Conversely, a solar eclipse can only happen at the other side of the lunar orbit, when the alignment is so perfect that the moon obscures the sun from our view. It takes about 14 days to get from full moon, when a lunar eclipse is possible at night, to a solar eclipse in the daytime. A solar eclipse the same day as a lunar eclipse is thus impossible, contrary to some descriptions of eclipses in literature, or interpretations of the crucifixion of Jesus of Nazareth, of which more later.

An example of this confusion is in the first edition of *King Solomon's Mines*. On p.149 the author, Rider Haggard, invokes a total solar eclipse at midday, yet during the subsequent night (p.164) his heroes are preparing for battle 'by moonlight'. In reality, of course, the moon would still be very close to the line of the sun, hence in daylight on the far side of the globe. That African night would have been pitch dark. He also required the totality track to run north to south from the United Kingdom to southern Africa, which is impossible. In later editions Rider Haggard changed the event into a lunar eclipse.

For now, I return to the central question: why do these eclipses not happen on a monthly cycle, like the waxing and waning of the moon itself?

As the earth spins, the line of sight of the sun tracks across the sky. Its path is called the ecliptic (see Figure 1C). Imagine this page as the plane of the ecliptic. If the orbit of the moon also lay on this same sheet, there would be an eclipse of moon or sun every fortnight. However, in reality the lunar orbit is inclined by a few degrees. This means that the moon will be above the sheet for about half the time, and below it for the other half. Thus it passes through the ecliptic—the sheet of the page—twice every month, once when travelling downwards and once, half a month later, when travelling upwards. The moments of traversing the ecliptic are called the 'nodes' (there being 'no d(isplacement)' between the pair). Thus a total eclipse can only happen when two things

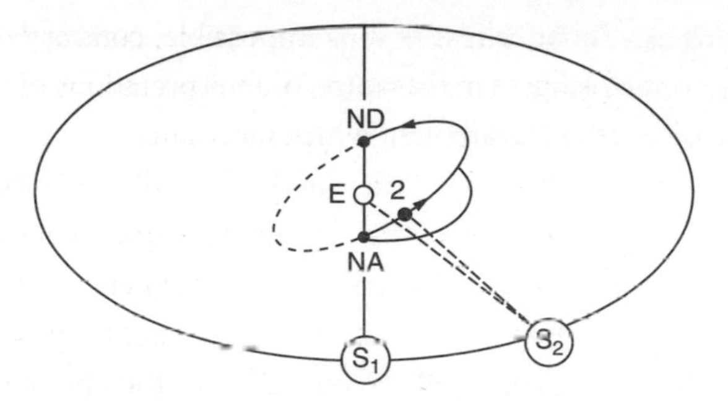

Figure 1C. The ecliptic and eclipses. From our perspective on earth, E, the sun and the moon appear to orbit us. The large ellipse represents the sun's orbit, which is in the plane of the page. The lunar orbit, the smaller ellipse, is oriented about 5° out of the page. If sun and moon are aligned at S1, when the moon is at NA earthlings see a solar eclipse, whereas if the moon is at ND there is a lunar eclipse. If the alignment is when the sun is at S2 and the moon at position 2, the moon's shadow goes above the earth; there is a new moon but there is no eclipse.

coincide: the monthly full moon or new moon occurs at the same time as the lunar orbit passes through a node.

This coincidence happens about once every 18 orbits, on average. So about half a dozen times every decade the moon's shadow hits part of the earth's surface. As the moon and earth continue in their orbits, this alignment survives for only a handful of hours, after which the moon's shadow penetrates empty space. But during those hours the earth has spun about a quarter of a revolution, so the hundred mile spot of darkness will have raced some five or six thousand miles across the globe. Thus it is that eclipse chasers located at a fixed spot on the earth's surface are spun into the shadow and out again within a few minutes. Only those situated along this narrow track will experience totality; outside this band there is a much larger area, which covers perhaps two thousand miles from north to south, where the sun appears to be partially eclipsed.

* * * * *

In the years after that memorable midsummer afternoon at school, solar eclipses came around, but in Peterborough they were always partial. I wondered if someone, somewhere else, in the far-flung pink reaches of the Empire, was watching totality. I was at a football match, once, sometime in the 1970s, sitting in the stand as Peterborough United played, and a partial eclipse of the sun was taking place above the tiered ranks on the opposite side of the field. But apart from me, no one seemed to have noticed.

Mr Laxton had one final message for us in that midsummer of 1954. As a youngster, in 1927, he had watched a total solar eclipse, which had crossed the north of England. He promised us that if we were ever fortunate to share such an experience, one would not be enough because 'After the thrill of the total eclipse you can't wait to do it again'. But 'wait you must' until that moment returns, and then make sure that you are in just the right place, or you will miss it.

On 30 June 1954, the track passed through the Shetland Islands, to the north of Scotland and missed England entirely. Thus our teacher had never assuaged that wish to see totality for a second time. Instead he had shared his excitement of his own youthful memories with his young pupils, and passed a legacy to our generation.

'When will we be able to see one?' we asked.

'In England?' he responded: '1999'.

On 11 August 1999, he assured us, the track would head across the Atlantic, clip the toe of Cornwall, then cross the English Channel into France, Germany, and beyond. And although it was more than five of my then lifetimes in an indeterminate future, which for an eight year old was like an encounter with infinity, I planned to be there.

2

Waiting for Godot

When I made that vow in 1954, I had no real conception of what 1999 meant. In 1954, The Beatles did not yet exist—they were themselves still in school. In the future, their music would dominate my life as a teenager, my years at university, marriage, and the infancy of two daughters. None of these were in my 1954 forward-look. The Beatles' supernova would eventually expire, and later John Lennon would die. Even then, another 18 years would still have to pass before 1999 arrived.

The political world would also change beyond imagination. In 1954, the British Empire was still pink and proud on our school maps. The Suez debacle, the Winds of Change, and Rhodesia's Unilateral Declaration of Independence were soon to change all that. The Berlin Wall had yet to be built. It would divide East and West for nearly three decades. When it came down in 1989, my date with the eclipse was still a decade away.

In June 1954, the father of modern computers, Alan Turing, died. The inventor of the World Wide Web, Tim Berners-Lee, was not yet born. By 1999, his invention would already be ten years old, and the means by which many of us would plan our expedition.

And throughout all these years, eclipses came and went, more regular than clockwork. None of them came my way, nor did I go seek them, but in the meantime I kept note of their passage.

The first was almost exactly one year later, on 20 June 1955. You would have to have gone to the South Pacific to see it, and someone obviously had because there were photos of the eclipsed sun in the newspapers. (Tim Berners-Lee was by then eight days old. Pictures of the 1999 eclipse would be downloaded from the Web within moments of the event. The word 'downloaded' was itself still decades in the future.). News of the 1955 eclipse would have passed me by but for Mr Laxton having awakened my interest.

The Pacific was again the location for the next one, on 8 June 1956. Now that I had begun to look out for them, it seemed that eclipses in the Pacific occurred at regular intervals. In 1957, however, intrepid explorers would have had to go to Antarctica in October. Then it was back to the Pacific once more, in October 1958. With the exception of an eclipse that crossed the southern part of Europe on 15 February 1962, every total solar eclipse from 1955 up to 1970 crossed the Pacific Ocean, either in whole or in part of its track.

The Pacific Ocean covers almost one half of the earth's surface, as becomes clear if you orient a globe such that the Hawaiian Islands are in the centre of vision. Viewed from space the earth would appear like aquamarine, almost the whole of one hemisphere coloured Pacific blue. If the moon's shadow hits the earth, then simple chance makes it likely that a random eclipse track will include the Pacific at some point. From my perspective in the 1950s, however, this seemed to confirm what I had learned in 1954: solar eclipses tend to happen in exotic locations.

I was intrigued also that solar eclipses seemed to recur just short of a year apart. My awakening had been on 30 June in 1954. This was followed by the eclipse of 20 June in 1955, with the next on 8 June in 1956. This regularity gave out after three successes, so I soon forgot it but it is a real phenomenon and not simply a boy's obsession with numerical patterns, as we shall see later.

* * * * *

Mr Laxton's impromptu description of solar eclipses in 1954 had first inspired my interest in science. I did not take up astronomy as a career, however, because by 1967 I was seduced by the quest to identify the basic building blocks of everything, and so became a high-energy particle physicist. In September 1970, having completed my Ph.D. I moved to Stanford in California, at the edge of the Pacific Ocean, on a postdoctoral fellowship for two years.

At least California borders the Pacific Ocean, but no total solar eclipses were scheduled in the vicinity of California during my visit. Eclipses were a part of astronomy, not particle physics, and thus were far from my professional life. I had no likelihood of going to the vastness of the Southern Ocean, where an eclipse seemed to happen almost every other year, and 1999 at home in England was still far in the future. Also, by that time, eclipses had slipped to the back of my mind.

Everything would change, however, one balmy Californian evening in February 1971. That night, I had a vision.

February in England is a grey, damp, cheerless month. For an English native in California, on the other hand, February is a delight. That year, the days were clear, and the air still; by night, once away from the city lights, the sky would be full of stars. February 11 had been one of those perfect days, and that evening after sunset I went out onto the balcony of my apartment to enjoy the full moon. But it wasn't there. Where the moon should have been, it appeared that someone had suspended an orange, without any visible support. The orange glowed as if there was light inside it, like a pumpkin at Hallowe'en, and it appeared to be hovering just out of reach above the roof of the next apartment.

To this day, I recall the shock of seeing this bizarre illusion. The apparition of an orange, which appeared to be the size of a fist at arm's length, was in reality a gargantuan moon some 240,000 miles away. The reason it wasn't the silvery lifeless colour of a normal full moon was because it was bathed in golden sunlight—golden because the sun's rays had passed through the earth's atmosphere and been reddened like we see at sunset. I didn't realise immediately, but I was watching a total eclipse of the moon (see plate 4 for the orange moon, or figure 2A in monochrome).

I was not aware in advance that this lunar eclipse was due to happen, let alone how stunning the apparition would be. Which begs the question: if the moon was totally eclipsed, how could it be visible, and shining like a luminescent orange? Something appears to be missing in Mr Laxton's demonstration in which the football earth cast its shadow totally over the lunar cricket ball. What does the real earth have that the football lacks?

The answer is that although the earth sends its shadow deep into space, like the football in that long-ago classroom, the earth's atmosphere plays a leading role. This thin blanket of gas bends sunlight around the surface as if the earth were a convex lens. Had one of the Apollo astronauts been standing on the surface of the

Figure 2A. Orange moon. The appearance of an orange moon during a lunar eclipse. (Photo courtesy of NASA.) See Plate 4 for a colour version.

moon at that moment, he would have seen the sun 'setting' around the full circumference of the distant earth, its light tinged deep orange, much as we earthlings see at a normal sunset. So, instead of a solid earth projecting only its dark shadow deep into space, its atmospheric lens brings sunbeams to a distant focus, where they brighten some of the gloom. Paradoxically then, a 'total eclipse' of the moon only darkens its centre; rays of ruddy sunlight, which have passed through the earth's atmosphere, illuminate its outer regions. The resulting combination of darkness and orange light gives an illusion of the moon in three dimensions, which the eye perceives as a small nearby sphere rather than a remote flat plate.

That night was doubly memorable. A few hours after the lunar eclipse, a major earthquake hit Los Angeles. Was it due to the eclipse? Probably not. There is no real correlation between eclipses and significant earthquakes. Sun, moon, and earth are nearly aligned twice a month, so the extra perfection that leads to an eclipse is, of itself, hardly the straw that breaks the back of the camel. Eclipses are magical, however, and if one coincides with some other dramatic event, people suspect there is some connection.

There are countless earthquakes without an eclipse, and large numbers of eclipses without major earthquakes. But both phenomena capture the imagination and over time become falsely linked in folk-memory, much as the coincidence of these two singular events has become locked in my own. A thousand years ago, such a coincidence would have become the stuff of legend, and added to the beliefs that inspired Shakespeare to identify: 'The pale-faced moon looks bloody on the earth' as one of the 'signs [that] forerun the death or fall of kings'.[1] In *Antony and Cleopatra,* he repeats this theme: 'our terrene moon is now eclipsed; and it portends alone the fall of Antony.'

William Shakespeare's fascination with both lunar and solar eclipses come together in *King Lear,* where the Earl of Gloucester declares: 'These late eclipses of the sun and moon portend no good to us.'[2] The play's first performance was in December 1606. In the previous decade there had been a memorable series of eclipses, visible in England, which undoubtedly inspired the playwright.

[1] *Richard II,* 7–10,15

[2] *King Lear,* I.2 112–113

England was remarkably lucky at that time. In 1598 there was a total solar eclipse visible over Cornwall, Wales, and North-West England, which was 98% of totality over Stratford-upon-Avon, while in July 1600 a total eclipse in Spain would have been up to 90% in England. In 1590 and 1601 there were annular eclipses, where all but a thin ring of the sun, or annulus, is obscured, both visible in England. Their psychological effect would have been comparable to complete totality nonetheless. There were also partial lunar eclipses in 1598 and 1601, visible in Stratford.

The total solar eclipse of 7 March 1598 would have been inspirational, additionally so if Shakespeare experienced its herald, a fortnight before: a 95% lunar eclipse in the pre-dawn of 21 February. The solar eclipse of 10 July 1600 was followed a fortnight later by a partial lunar eclipse, once again in the pre-dawn. The climax of these eclipses came in 1605, however, when a trio of dramatic eclipses occurred within just a few months.

On 3 April 1605 a lunar eclipse occurred at sunset with 99% of the moon in shadow. This was the largest partial lunar eclipse of the seventeenth century. The moon was near perigee, meaning at its closest to earth, which would have made it appear relatively big, and the resulting eclipse, which continued for more than an hour after sunset, would have been memorable. This was trumped by events in the autumn, however. On 27 September, another lunar eclipse took place. This lasted for three hours, in the pre-dawn, and presaged the climax of the sequence as, two weeks later on 12 October, the sun was eclipsed in the middle of the day. This was total in the southwest of France, and over 95% was blacked out in Shakespeare's Stratford. He would have seen the sky darken, animals become quiet, and birds begin to roost.

So astronomy enters the debate on when Shakespeare wrote *King Lear* (Figure 2B). Literature scholars have determined its creation to sometime between 1603 and 1606, which is consistent with his reference to these 'late eclipses' being the 1605 trio. This doesn't necessarily enable us to date his play to 1605, however. The eclipse record suggests that the inspiration for Gloucester's remarks could well have originated in the eclipses of 1598 and 1600, and that the trio of eclipses in 1605 brought Shakespeare's creation to its climax. That the collection of eclipses from 1598 to 1605 inspired the reference seems beyond dispute.

The conjunction of lunar and solar eclipses, just 15 days apart, in 1598, 1600, and again in 1605, was no coincidence. Instead, they are classic examples of the planes of sun, earth, and moon having been so well aligned that the moon, having cast its shadow onto the earth for the solar eclipse, was also in the Earth's shadow half a lunar orbit earlier or later.

Such conjunctions could be one of the possible clues that ancient astrologers used to 'predict' an eclipse. In any event, it is a model example of the minimum time possible between a solar and lunar eclipse; they cannot occur on the same day, contrary to some interpretations of the events that accompanied the crucifixion of Jesus of Nazareth, such as in Acts 2:20: 'The sun shall be turned into darkness and the moon into blood.'[3]

Three gospels tell the same story of the crucifixion, and record that 'the sun was darkened' in the middle of the day. In Luke's gospel, as presented in *The New English Bible*: 'by now it was about

[3] See http://scienceandapologetics.org/engl/g19.html for discussion of Luke's gospel and reference to Acts of the Apostles, and the conflation of prophesies about astronomical events and their application to the crucifixion.

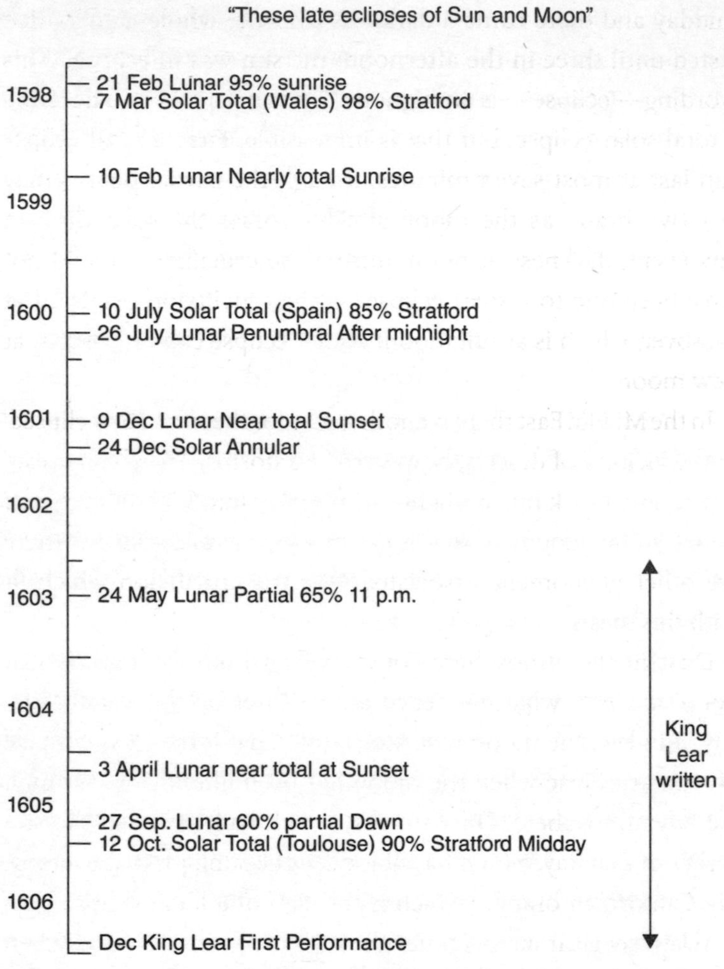

Figure 2B. Timeline of eclipses and Shakespeare's *King Lear*. Details of the solar eclipses and their paths can be found at http://eclipse.gsfc.nasa.gov/SEcat5/SE1501-1600.html and http://eclipse.gsfc.nasa.gov/SEcat5/SE1601-1700.html. The lunar eclipse information is at the corresponding sites but with LE (lunar eclipse) in place of SE.

midday and there came a darkness over the whole land, which lasted until three in the afternoon; the sun was in eclipse.' This wording—'eclipse'—is ambiguous. It has been taken to describe a total solar eclipse, but that is impossible. First, a total eclipse can last at most seven minutes, though the partial phases may last two hours as the moon slowly crosses the solar disc. In any event, darkness at noon, during the crucifixion, could not have been due to a solar eclipse, as the crucifixion occurred at Passover, which is at full moon: a solar eclipse can only occur at new moon.

In the Middle East there is another way for the sun to be 'eclipsed' in the vicinity of deserts, however: dust storms. These can create havoc, and block out sunlight, turning day into night, for several hours. In my judgment that is the probable explanation, as there are other phenomena associated with the crucifixion which fit with this thesis.

Dust in the atmosphere could explain both the darkness at noon and also what happened after sunset on that fateful Friday, if indeed the moon rose blood-red. That is the classic image of a lunar eclipse when the moon has been illuminated through a dusty atmosphere. Dust storms, which had obscured the sun earlier in that day, would have blooded the sunlight that evening. My Californian orange, which is the sign of a lunar eclipse seen in relatively clear air, becomes the colour of venous blood when illuminated through a polluted atmosphere.

Everything is consistent with the crucifixion having coincided with a total eclipse of the moon. Furthermore, Passover takes place at full moon, which is the very time of month when a lunar eclipse can occur. And the clincher: in 33 AD, Passover was on

Friday 3 April, and the astronomical records reveal that the moon rose that evening, eclipsed.

This bloody moon would have been a dramatic sight. The coincidence of Christ's death with the moon being darkened to the colour of his blood would have made both a powerful memory and a compelling metaphor for his followers. Several decades passed before the gospels were written. Memory is fickle, and the gospels were created from folklore that had been passed from one generation to the next. Thus false memory of another solar eclipse might explain the allusion to a solar eclipse as the cause of the darkness at noon. And indeed, there had been a total solar eclipse visible in the region, but not on the day of the crucifixion.

Three years earlier, on 24 November in 29 AD, the path of totality had traversed Damascus and Beirut, and passed north of Jerusalem. Given the psychological shock that this event would have had on the population, one might expect to have found it mentioned in the gospels: at a time when Jesus' ministry and miraculous illusions were in their pomp, a total solar eclipse would be a godsend. How come that record of this eclipse has been—eclipsed? A possible explanation is that it *is* mentioned: it has become conflated in the folk memory with the darkness of the noonday sun at the crucifixion, three and a half years later. In any event, we can be sure of one thing: thanks to astronomy, we can date the crucifixion to Friday 3 April in 33 AD.

* * * * *

The propaganda benefit of associating eclipses with singular events touches our psyche profoundly. Primitive societies could hunt at night when the landscape was illuminated by the full moon. However, a lunar eclipse, which snuffed out moonlight for several hours at the very time when the moon was expected to be at its brightest, would be a considerable worry. That the moon also turns to the colour of blood would compound their fear.

Ancient hunters watched the stars. Dark nights would have revealed the Milky Way. The night sky was familiar to them, unlike the experience of modern city dwellers. The ancients would have noticed shooting stars, occasional comets, and lunar eclipses. The moon is totally eclipsed on the average twice every three years. The event lasts for several hours and can be seen by everyone on the night-side of the earth, weather permitting. Even under cloud, the darkening would be apparent. In addition to these total eclipses there are partial eclipses of the moon, which happen about every eight months. The average lifespan of a human was less than it is today, but even so, more than a score of lunar eclipses—both total and partial—would happen in a lifetime. The ancients would have taken note of these nocturnal events, and eventually discerned a pattern—of which more later.

Solar eclipses would have been another matter. A total eclipse of the sun would be truly terrific, and terrifying. Totality is rare, and most people pass their whole lives without seeing one. If a total eclipse of the sun occurred in the homelands of a tribe, it would be a singular event, long remembered and passed down in folkore. Partial solar eclipses, however, might occur half a dozen times in a lifespan. Most of those partial eclipses won't dim the light, and might escape notice entirely other than by sages, witch-doctors and astrologers. Total eclipses, by contrast, would create fear and panic when the source of light, heat, and life itself was suddenly blotted out. Minutes later there would be relief as daylight returned. Reprieved by the gods, the audience would be ripe for exploitation by charlatans, or by those 'in the know'.

Foreknowledge of an eclipse would thus give immense power to the informed. This has been a favourite theme in literature, but also has basis in fact, as in the case of the explorer, Christopher Columbus.

Columbus made four voyages to the New World, on the final one of which he ran into trouble. He left Spain with four ships, but a plague of shipworms ate holes in the wood. He abandoned two of the craft, and eventually in June 1503 he had to beach the remaining pair on what is now known as Jamaica.

The native Arawak Indians were initially friendly, and supplied the Spaniards with food in return for trinkets, such as mirrors and other goods, which were trashy to the Europeans but novelties to the locals. Time passed, and tensions mounted. Half of his crew mutinied, murdered some Indians, and escaped the island in homemade canoes. For the Indians, the novelty of sixteenth century euro-trash had by now worn off. They ceased to supply any food, and Columbus' remaining crew faced starvation.

Like any competent sea captain, Columbus carried an almanac, which contained star-charts to aid navigation. He noticed from its tables that on the evening of Thursday 29 February 1504 there would be a total eclipse of the moon, soon after moonrise. He decided to use this to pressure the Arawak chieftain into helping him.

Columbus arranged a meeting with the chief, three days before the eclipse was due. He explained that he and his men worshipped a powerful God, who was very displeased that the Indians were not providing succour. To demonstrate his wrath, the God would darken the moon when it rose three evenings hence. He would redden it, like blood, to signify the destruction that He would bring if the Indians failed to help Columbus.

That Thursday night, when the sun sank beneath the western horizon and the moon rose in the east, almost at once it was obvious that something was wrong. The lower edge of the moon was dark, as if a piece of the orb had been removed. As the moon rose, and twilight gave way to darkness, instead of a silvery hue, the moon had a dark, dirty reddened appearance. Within another hour, instead of the usual bright full moon, a dim red ball hung in the air above the horizon.

Columbus' log reported that the Arawaks ran to his ships from every direction 'with howling and lamentation'. They brought provisions and pleaded with Columbus to intercede with his God on their behalf in return for them donating the goods. Columbus now played clever. As the moon would remain eclipsed for another hour or more, Columbus retired to his cabin in order, he said, to plead with his God on their behalf.

Back on ship, he timed the eclipse with an hourglass. A few minutes before totality was due to end, Columbus emerged from his cabin. He announced that his God had pardoned the Arawaks and would now allow the moon to return. Columbus had timed it well. The moon duly reappeared, and the Arawaks kept his men supplied with provisions for another four months until a rescue ship arrived from Spain to collect Columbus and his stranded crew.

* * * * *

Columbus had deliberately used an eclipse to his advantage. Historically, the awesome phenomena were interpreted as omens. Major events in history, such as the deaths of kings and the outcome of battles, were often linked with eclipses in the folk-memory. There is no physical causation: alignments of sun and moon exert no special cosmic forces. However, the psychological impact of such events can act as self-fulfilling prophecies. As an inferior tennis player when filled with self-belief may defeat a superior opponent, so eclipses can be used to help or to hinder performance.

For example, in 413 BC the Athenians and Spartans fought the Battle of Syracuse. The Athenians were encamped, ready to attack. The states of the respective armies suggest that the Athenians would have held the advantage, and would have been victorious if they had launched an assault there and then. However, a lunar eclipse occurred. Astrologers declared this to be an unlucky omen. The Athenian commander duly decided it was not a favourable time to attack, and so held off for a month. This handed the advantage to the enemy, who duly routed the Athenians.

The earliest story linked to an eclipse goes back to China in the Hsia dynasty, possibly 2110 or 2137 BC. The Chinese believed a solar eclipse to be the result of a dragon eating the sun. Astronomers were charged to predict when an eclipse was due, so that crowds could be organised to make noise, beat drums, and fire arrows into the air to drive the dragon away. Unfortunately the Astronomers Royal, Hsi and Ho, failed to predict the eclipse. Their incompetence angered the Emperor who ordered their execution.

That, at least, is the story, but it seems to me apocryphal. In some versions of the tale Hsi and Ho were supposedly drunk, hence their failure. A more sober interpretation of this tale might take into account that Hsi-Ho—'brightness and harmony'—was the Chinese name for the sun goddess. In ancient mythology there were 12 moons and ten suns; Hsi-Ho controlled the time when each sun took its turn in lighting the world. A total disappearance of the sun, which is how a total solar eclipse would appear, equates to a failure by Hsi-Ho. It was the sun goddess who was asleep, not two human astronomers, though it is quite possible that they were beheaded, whatever their names might have been.[4]

* * * * *

Ancient Chinese and Greek astronomers routinely measured the motion of the moon, planets, and stars. They were aware that the moon changes speed as it orbits the earth, as does the earth in its circuit of the sun. By the first century BC, Greek astronomers were apparently able to predict a solar eclipse to within an accuracy of about a month.

They could probably have predicted lunar eclipses much earlier as this required less precision. The reason is that the earth's shadow extends more than 8000 miles across, whereas the moon's shadow on earth at totality is of the order of only one hundred miles, and the breadth of totality even less. At the distance of the moon, this relatively small extent subtends an angle of less than 1/30 of a second of arc. So to predict a solar eclipse requires knowing the moon's orbit to this level of accuracy.

By 1620 Johannes Kepler had discovered the laws of planetary motion, which established that their orbits around the sun are ellipses with the sun at one of the two foci, and that a line segment joining the planet and the sun sweeps out equal areas in equal time

[4] C.R. Coulter and P.Turner, *Encyclopaedia of Ancient Deities*, Routledge, 1999.

intervals. This helped inspire Isaac Newton to his law of universal gravitation. Falling apples and the moon feel the gravitational tug of the earth; in turn all of these, as well as the planets and comets, feel a universal gravitational attraction to the sun. These theoretical breakthroughs were accompanied by precision astronomical measurements, in particular by the Astronomer Royal, John Flamsteed.

Flamsteed used tables of data on the motion of the moon and the earth through the heavens to, predict the solar eclipses of 1666 and 1668.[5] Following Newton's elucidation of the laws of gravity in 1687, eclipse prediction changed from empirical rule of thumb to an exact science thanks to Edmond Halley. Halley used the celestial dynamics of Isaac Newton to determine not just the motion of comets, but also the chronology of eclipses. He analysed the motion of the moon, its path curving up and down relative to the plane of the earth's orbit about the sun and calculated the timing of a total solar eclipse over London in 1715.

Halley's prediction was correct to within 4 minutes, and his map of the eclipse path was accurate to within just twenty miles. After the eclipse he refined his calculation to take account of the discrepancies, and updated his prediction for the eclipse of 1724. His revised diagram of the paths of these two eclipses literally shows the state of the art of eclipse science in the early eighteenth century.

Today, eclipse maps can factor in the precise shape of the moon's profile as measured by laser altimeters on lunar orbiters. This is a

[5] See p26 in *An Account of the Rev John Flamsteed*, the first Astronomer Royal, by Francis Baily, Admiralty, London, 1835.

remarkable testament to the power of modern computation. Halley's 1715 eclipse chart (Figure 2C) was invaluable in its time; today we can watch a dynamic illustration of the path of that historical eclipse on a laptop.[6]

* * * * *

Since Edmond Halley discovered how to compute the locations and timings of eclipses accurately, it has become possible to back-calculate their chronology. This enables us to identify the dates for famous eclipses, and thereby set markers for historians. For example, eclipses are mentioned in the Old Testament of the Bible, such as in Genesis when Abraham is in Canaan and: 'when the sun was going down, great darkness fell upon him'. The location and timing fit with a solar eclipse that occurred about 6.30 p.m., local time, on what would have been the 9 May in 1533 BC.

Another famous biblical eclipse is recorded in the book of Joshua. This can be dated as 30 September 1131 BC. Its most remarkable feature is that this eclipse included a miracle, as the sun and moon apparently stopped and even reversed their motion. The Old Testament scholar, Robert Dick Wilson translated Joshua 10:12b-14 as follows: 'Be eclipsed, O sun, in Gibeon, and thou moon in the valley of Aijalon! And the sun was eclipsed and the moon turned back, while the nation was avenged on its enemies.'[7]

The eclipse certainly took place, but the sun does not stop, nor can the moon turn back in its tracks at any time. I decided that

[6] For an animation of the 1715 eclipse, see https://en.wikipedia.org/wiki/Solar_eclipse_of_May_3,_1715#/media/File:Solar_eclipse_of_1715_May_3-animation.gif

[7] See https://www.blueletterbible.org/faq/don_stewart/don_stewart_625.cfm

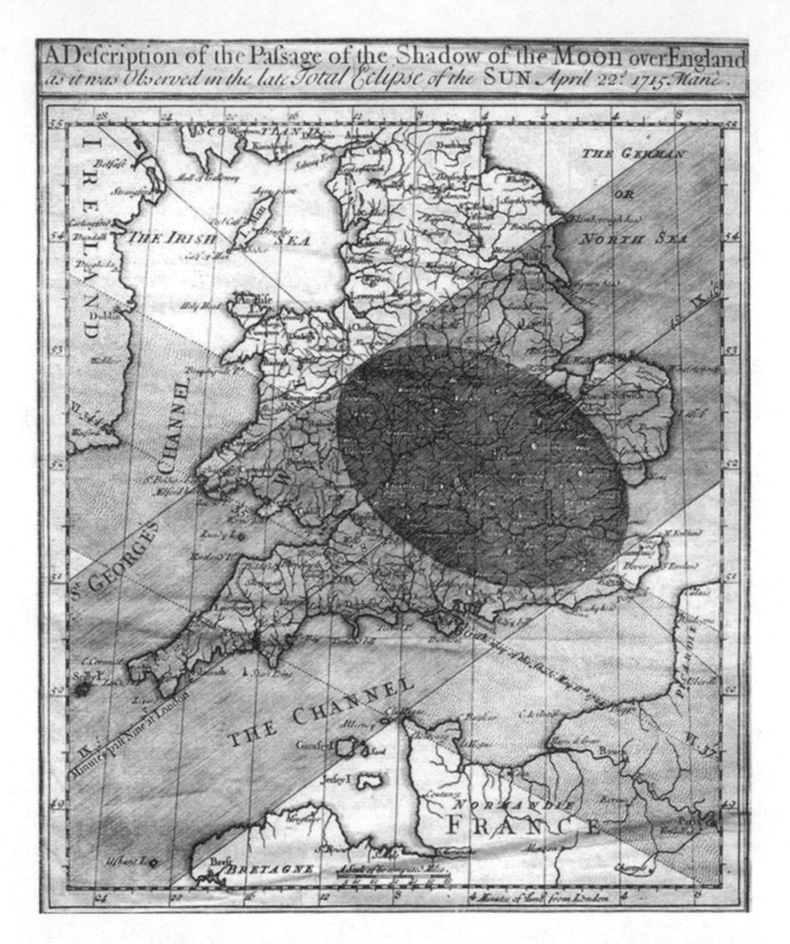

Figure 2C. Halley's eclipse chart of 1715. After the 1715 eclipse, Halley improved his calculations and redrew the path of the eclipse; this is the dark area from the bottom left to the top right, which includes an outline of the shape of the moon's shadow. He also predicted the track of the 1724 eclipse, which is the band from the mid-left to the bottom right. South Wales and much of southern England thus experienced totality twice in just nine years. (Courtesy of the University of Cambridge Institute of Astronomy Library) (Source: http://www.phenomena.org.uk/eclipses/eclipses/totaleclipses.html)

this must be some optical illusion associated with a total eclipse, but nowhere could I find an explanation. I made a note to watch for it when, eventually, I saw one for myself, and try to solve the mystery.

The earthquake that destroyed Solomon's Temple in Jerusalem occurred around the time of a solar eclipse. Back calculation of the timing of past eclipses confirms that a total eclipse of the sun was visible in that region on the 15 June 763 BC. This biblical eclipse features in Amos 8.9: 'I will cause the sun to go down at noon and I will darken the earth in the clear day.' The chronicles also record that a bright comet was seen around that time. Halley's Comet, which reappears with a period of some 76 years, would have been visible in August that year, and also have been brighter historically than today. Although we have no direct proof that the comet referred to was indeed that identified by Halley, it is intriguing, and is also an example of how the clockwork of astronomy constrains the matrix of possibility.

The link with Halley raises a question, however. As he only determined the mechanical chronologies of comets and eclipses around 1700, how were eclipses predicted in earlier epochs? How did the author of Columbus's almanac provide such accurate and valuable information? The answer, in part, is that eclipses, like the tides, come and go to a rhythm. And it can be possible to discover the pattern without having any understanding of why it comes about.

* * * * *

You can predict when a high tide is likely to occur without needing to solve technical equations. Nor is it necessary even to know the principles that lead to the rise and fall of the oceans. But to illustrate how, it's helpful to explain first something about how the tides arise.

The moon exerts a pull on the earth through the force of gravity. The strength of the gravitational force between two massive objects dies away in proportion to the square of the distance between them. Thus the moon tugs the surface waters of the earth, tonne for tonne, more powerfully than it pulls at the more remote bulk of the planet itself. This lifts the water on the moon-side of the earth, and forms a tidal lump. Meanwhile, the earth is tugged more strongly than are the waters on its far side. The result is that two bulges of water, one beneath the moon and the other across the globe, follow the moon as the earth spins in its daily round.

One day later, when the earth has spun through a complete revolution and the moon is in the same relative position, the high tide should re-occur. This would be 24 hours later but for the fact that the moon is itself orbiting the earth once every month. It takes about an hour to complete this 'extra' four per cent of its motion, and as this is in the same direction as the earth's spin, the high tide tomorrow will be about an hour later than it was today.

This is a fairly good approximation to what actually happens. Even if we had no idea of this cause, we could deduce the pattern. After just a few days it would become obvious that the high tide is progressively delayed. By keeping a record of the times of the daily tides, and over a year also noting how their relative magnitudes rise and fall with the seasons, an almanac of tides could be constructed. Our pedagogic description of the 'why' and 'how' is not needed if all one wants is a timetable of 'when'.

As it was for tides, so it is for eclipses. A major difference, however, is that it takes longer to repeat the relative positions of sun, earth, and moon required for eclipses. These are 19 years, in some cases, and as a result are harder to recognise. But in societies where records have accumulated over many decades, the code is eventually revealed, and has been re-discovered independently by many cultures. Before 2000 BC, the Chinese, for example, had identified some patterns (Figure 2D), though not all as the unfortunate astronomers Hsi and Ho supposedly discovered. The Babylonians in the seventh century BC, Egyptians and the ancient Greeks two to four hundred years later, and the author of Columbus's almanac, had all decoded the eclipse repetition. Indeed, when modern eclipse chasers use their jargon and say the eclipse is part of 'Saros number N', they are using the Greek word saros, which means repetition.

It is easy to see that eclipses are not random events. Here is a list of the dates of total solar eclipses in the two decades from my virginal British eclipse of August 1999, to the great American one of August 2017. Written as day/month/year they are:

11/8/99; 21/06/01;
4/12/02; 23/11/03;
8/4/05; 29/3/06;
1/8/08; 22/7/09; 11/7/10;
13/11/12; 3/11/13;
20/3/15; 9/3/16;
21/8/17

Starting with the eclipse in December 2002 until that in 2016, each solar eclipse is partnered with one that occurs almost a year

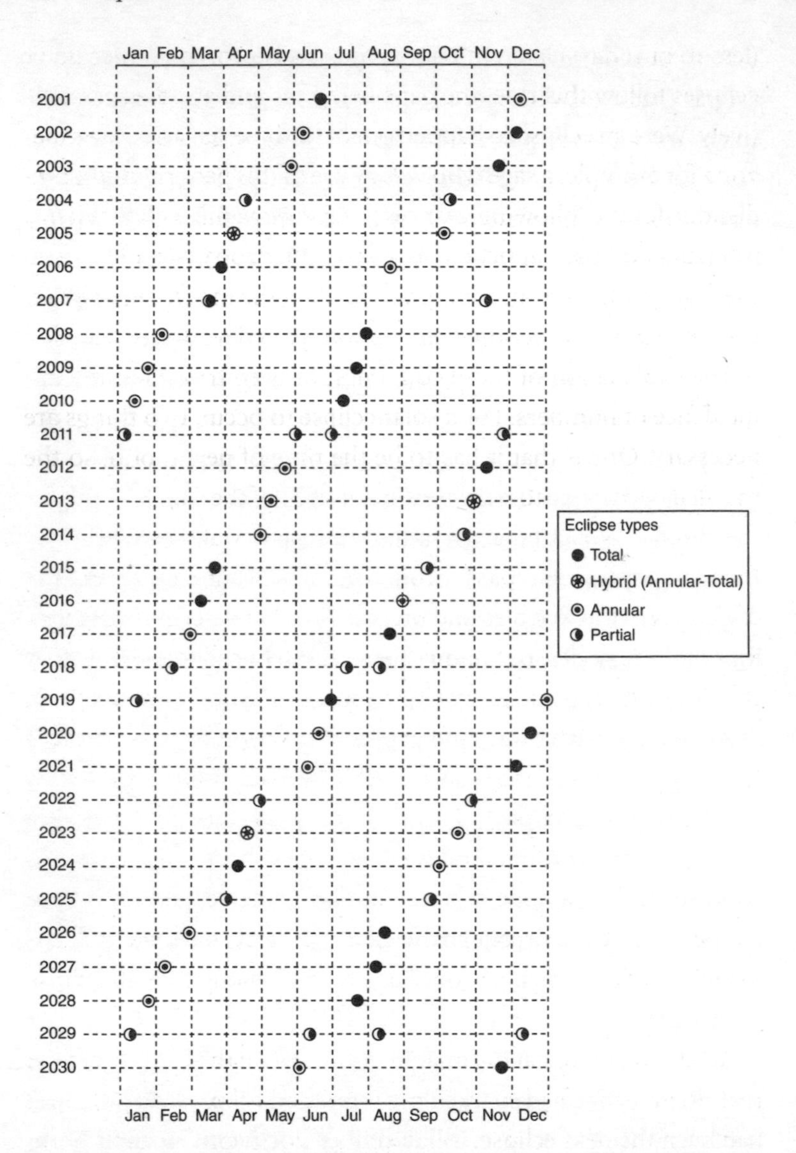

Figure 2D. Solar eclipse pattern. Note how the pattern of eclipses repeats on an eighteen years cycle, displaced by about ten days. (Adapted from Mark Littmann, Fred Espenak, & Ken Willcox, *Totality: Eclipses of the Sun* (Oxford: Oxford University Press, 2008), p. 15, 'Solar eclipses 2008-2026'.)

(less 10 or 11 days) later. In 2008, 2009, and 2010, three successive eclipses follow this rule, the gaps being 355 and 354 days, respectively. Were an eclipse to happen out of the blue, as on 4 December 2002 for example, a sage who was aware of this pattern might predict one for the following year on 22 or 23 November. And when a total lunar eclipse happened on 9 November 2003, as it did in fact, the sage would have an additional clue that the moon might be on course to blot out the sun, half an orbit, or 14 days, later.

The explanation of these couplings, or even trios, lies in a coincidence of numbers. For a solar eclipse to occur, two things are necessary. One is that it has to be the time of new moon, so the moon has a chance to be in the line of sight of the sun. Second, for the moon to blot out the sun rather than passing above or below it, the lunar orbit has to pass through a node. I cannot compute the dynamics of the wobbles and precessions that determine the timings of these nodes, but I don't need to. All I need to know is that the time between successive new moons—'synodic months'—is 29.53 days, whereas the time gap for the moon to pass through successive nodes is 27.21 days. Twelve synodic months equate to 355 and one-quarter days. This is almost the same time as it takes to pass 13 times through nodes (a few hours short of 354 days). So if on some date a node and new moon coincide, and an eclipse happens, there is a decent chance that 354 or 355 days later the coincidence of alignments is once again good enough for an eclipse to take place.

At last I realised that my imagined patterns in the 1950s had been real. 30 June 1954 had sparked my interest in eclipses; 20 June 1955 had been the next eclipse, followed like clockwork by another on 8 June 1956. The apparent extension of the gap from ten days short of the year to 12 was due to two things. First, 1956 was a leap year, but for which the eclipse would have occurred on 9 June. Also, the eclipse was centred in the Pacific and in the vicinity of the date line. Further eclipses in October 1957 and again in the same month of 1958 merely confirmed the cycle for me, but now raised another question: why does the sequence terminate so soon? Here is the answer.

If the time gaps between 12 new moons and 13 nodes had coincided exactly, there would be eclipses every year on this 355 days cycle. If in addition this coincidence were to have amounted to an exact number of days, to the minute, the longitudinal range of the eclipse path would repeat also. In reality there's a mismatch.

First, the coincidence is not exact to the minute, so the earth has spun to a different orientation when the shadow hits. In other words, the shadow will span a different longitude each time. Second, totality doesn't require the moon to be precisely at a node. If, instead, it is slightly before or after a node, the inclination of its shadow will marginally differ. Thus the moon's shadow will hit our surface at different latitudes. For example, suppose that on the first occasion in a sequence the eclipse path followed the equator. Were the next alignment to occur just before or after the exact node, the shadow on this occasion would project high into the northern or southern hemispheres. Thus an eclipse could occur again, nearly a year later, though at a different latitude. Now move on one year more, and the deviation from the node will be yet greater. In this case the shadow will pass overhead the poles, and escape into space. The sequence of earthly eclipses will have ended after just two or three years.

A more precise match between synodic months and nodes takes place about every 18 years. 233 synodic months and 242 nodes agree within just 50 minutes every 18 years and 10 or 11 days (depending on whether four or five leap-years have elapsed in the meantime) and eight hours. This coincidence is so good that the moon's shadow can hit the earth on up to 80 occasions, over more than a millennium, before it overshoots one or other of the poles.

The great American eclipse of 21 August 2017 and my virginal one in England on 11 August 1999 are examples of such a saros, or repetition. In England the eclipse was at 11 a.m. and in the USA, on its west coast, the eclipse is also due at a similar hour that morning. This illustrates the effect of the eight hours, or one-third of a day, in the time gap. In that amount of time the earth has spun one-third of a rotation on its axis. The longitude of the path is thus displaced westwards by some eight time zones—from England to the west coast of the USA. The alignment of the node is not perfect, however, the latitude shifting by a few degrees on each successive occasion until eventually, on 17 April 3009, the final eclipse in this particular saros cycle will take place, over Antarctica. This is why the track in 2017 is displaced in both latitude—to the south—and longitude—west—relative to that in 1999.

This all applies to members of a single saros. At any time there is a number of independent saros cycles in operation, all intermingled. For example, in our list above, the 1999 eclipse is a member of what astronomers refer to as 'saros 145'. That of 2001 was in 'saros 127', the remaining dozen in the list scattered randomly without repetition between saros 120 and saros 146. Not until that of 21 August 2017 is there a repetition: 'saros 145', the same as that of the 1999 event. For future reference, the saros cycle 127 repeats on 2

July 2019, giving totality in a band across the South Pacific and South America. This illustrates that until the saros code had been cracked, eclipses would have appeared to be random: successive 'correlated' eclipses, which are members of the same saros, come around only rarely.

As the paths of totality are so narrow, and displaced so far across the globe on 18 year intervals, it would be impossible for the ancients to experience more than one total eclipse in a particular saros. So, how did they discover the patterns?

First, while it is indeed the case that nowhere is there any overlap in the paths of totality, the sun is partially eclipsed over a vast area. Whereas the width of the path of totality is typically less than 100 miles, a partial eclipse might extend for some 5000 miles. For example, in 2017, the associated partial eclipse spreads across the globe and will be visible as far away as England, where its forerunner in 1999 was observed in totality. Thus it is possible that the sequences were identified through a combination of partial rather than just total solar eclipses. More likely, however, is that ancient watchers of the heavens detected the patterns of lunar eclipses. Whereas a total solar eclipse lasts for a relatively short time, and is total only over a limited region, a lunar eclipse lasts for hours and is visible everywhere on the night-side of the earth.

This widespread visibility of a lunar eclipse enabled the British to measure the width of the Atlantic Ocean in 1584. They had colonised America, and settled on the east coast in what is now Virginia. They needed to know their exact coordinates of latitude and longitude, not least so that ships coming from London would be able to locate them. It was easy to measure latitude, the height above the equator, simply by measuring the angle of the sun at its maximum in the sky—noon—or by the angle of polaris, or other known stars at night. Measurement of longitude, however, was a conundrum in an era when accurate clocks had yet to be made. This is where an eclipse of the moon came to their aid.

On the night of 17–18 November 1584, a lunar eclipse was due. It would be visible in both London and Virginia. Unlike a solar eclipse, where the moon's shadow takes several hours to traverse the earth's surface, in a lunar eclipse the earth's shadow is projected onto the moon and is visible at the same instant from everywhere on the night-side of earth.

The American settlers thus prepared a pendulum clock, which they synchronised with the sun at high noon—their noon. When the lunar eclipse began, they noted the time as it appeared to them. An analogous set of measurements was made at Greenwich in London. Thus at the same instant, as determined by their common view of the eclipsed moon, the two clocks recorded the respective local times. The difference between the hour, as measured in Virginia, and that in Greenwich indicates, in modern parlance, how many time zones separates the two. More precisely, it determined the longitude relative to Greenwich. In effect, it was by observing a lunar eclipse that the width of the Atlantic Ocean was first measured in 1584.

* * * * *

As we saw earlier, the work of Edmund Halley in the eighteenth century gave the chronologies of both solar and lunar eclipses a fundamental scientific explanation. With this accurate natural chronometer, it became possible to calibrate the passages of time with unprecedented accuracy. Not only could astronomers predict future eclipses, but also by back-calculation they could determine the dates and locations of eclipses in history. Thus it was possible to date events in the bible, as we have seen, and contribute to literary debate about the completion of Shakespeare's *King Lear*. It has enabled historians to identify errors in ancient calendars, by the simple expedient of comparing historical records of eclipses with the astronomical events as computed from the rules of celestial mechanics.

Thus, for example, eclipses expose the haphazard nature of the calendars of Roman Emperors, which bore no relation to the seasons. Instead, emperors had gerrymandered the calendar by arbitrarily altering the lengths of the year in order to maximise their earnings from annual tax revenues. The Roman chronicler Livy recorded that a lunar eclipse occurred on the eve of a battle, which took place, according to the calendar, on 3 September. Far from this being in early autumn, however, this lunar eclipse occurred in what we call June, near midsummer. The calendar was nearly two months in discord with the seasons. By 46 BC, Julius Caesar had to add 80 days to the calendar to rectify the machinations of former emperors.

The first total eclipse recorded in England was in 664 AD. Modern computations confirm this to have taken place on 1 May, and not on 3 May, contrary to the records kept at the time in English monasteries. Monastic archives in Ireland and in continental Europe, however, all agree that it was on 1 May, in accord with celestial dynamics. Monks of the time, such as the Venerable Bede, were careful: when they entered three, they meant three. The failure to enter the correct value—one—was deliberate. The celestial chronometer reveals that the record was falsified.

The reason for this deception may be associated with the Synod of Whitby, which took place that year. This was not a lunar synod, a meeting of the moon, but a 'coming together' of clerics, which is the most common contextual use of 'synod' today. The clerical synod decided to leave the Celtic church and join the Roman church, whose doctrines had been brought to the British Isles by Saint Augustine about 60 years earlier. This transfer of allegiance arose because King Oswy of Northumberland had been converted to Rome, not least because the Roman computation of the date of Easter, deemed seminal in the Christian doctrine, was supposedly superior to that of the Celts.

The Roman supporters further argued that the eclipse was a sign from God that the doctrines of the Celts were false. In reality, however, the Roman computation of Easter was wrong. According to their erroneous calendar, the date of the eclipse would have been on 3 May. When later they discovered the error, they tricked the Synod into believing that the eclipse had indeed occurred on the third, rather than two days earlier. This deceit was deliberately maintained in the false record.

* * * * *

My vision of the blood-orange moon in 1971 had been so remarkable that I now began to appreciate Mr Laxton's promise that it would only be if we saw a total solar eclipse that we would understand its seduction. If he had said that about a total lunar eclipse, then that sight in the California night would have proved him right.

Following that unexpected epiphany, my interest in eclipses reawakened. I had read about them, as in the several narratives above, and mused at how science can reveal truth and expose fictions in historical narratives. As I remarked before, one particular case from the biblical record intrigued me, namely the assertion in Joshua that the sun stopped and the moon 'turned back' at eclipse.

This is clearly physically impossible. Had it happened, it would have truly been a miracle. However, although I knew nothing as yet from personal experience about total solar eclipses, the wonder that they clearly evoked in spectators made me suspect that this backtracking might be some form of optical illusion. This I added to my reasons for wanting to see one for myself, with the goal of understanding Joshua's report.

As for the celestial dynamics that leads to eclipses, of sun or of moon, the most immediately obvious consequence is that 14 days later, when the linear positions of earth and moon have become interchanged, the conditions may still be ripe for an echo eclipse of moon or sun. That is what happened in the decade prior to the production of Shakespeare's *King Lear*, and that is what happened after my lunar eclipse in 1971. A fortnight later, the trio were still sufficiently well aligned that the sun was partially eclipsed, when viewed from Europe. Had that European event been a *total* solar eclipse it would indeed have been ironic that I was far from home, in California.

The following year a total solar eclipse occurred in North America, the path crossing eastern Canada on 10 July 1972. It was a 60% partial eclipse in England, and I saw it—in England. I had returned home from California just a week before. That irony of being out of phase with totality became a leitmotif for the next three decades as I was invited to visit exotic places, where total solar eclipses would occur, but never when I was there. I would not have the chance to test my theory about Joshua any time soon.

3

Preparation

My travels as a physicist in the final decades of the twentieth century took me to several places where total solar eclipses had happened, or would soon happen. Unlike Swiss railways, however, I was never present on time for the actual event. Not once did the eclipse schedule coincide with mine.

I spent a month at the University of Adelaide, in Australia, but too late to see the total solar eclipse, which traversed South Australia on 23 October 1976. I visited the University of Washington in Seattle, but sadly 20 weeks after the total eclipse of 26 February 1979. My only visit to Central America was six months before my hosts were darkened by the total eclipse of 11 July 1991, and I missed that in Brazil, on 3 November 1994, by just a few weeks.

I seemed destined to miss totality. I love cricket but missed the opportunity to watch England play India in the Golden Jubilee Test Match at Mumbai in February 1980, when proceedings were interrupted on the 16th due to a total eclipse of the sun. This is the only occasion in the history of cricket that 'eclipse stopped play' has been recorded. When I visited Rajasthan in the New Year of

1995, I discovered that I was ten months early for the total eclipse of 24 October. By then, having waited 40 years, a mere 40 months remained before the dream of Cornwall 1999 could come true, so I returned home rather than stay in India.

Meanwhile, some enterprising physicists in Vietnam organised a conference in Ho Chi Minh City (Saigon) during that October of 1995, just 150 miles from the path of totality. Unfortunately, I could not go. One of my colleagues, Michael Duff, unaware of the impending solar eclipse, was present. He told me what happened on eclipse day.

> The organisers arranged for four buses to take the conference participants to watch the total eclipse near the Cambodian border. We were supplied with conference baseball caps and special Mylar glasses to protect our eyes. Our bus, which was carrying Nobel Laureate Norman Ramsey, became separated from the others. It was soon obvious that the driver was lost.

This was all before the era of ubiquitous mobile phones. The conference included some astrophysicists, and one or two of them had brought GPS devices on the trip. They worked out the shortest route to the twilight zone and directed the driver to the spot where there would be complete totality. At the appointed time the bus stopped and everyone got out. Other than that they were in the path of the eclipse, no one had any idea of where they were. It happened that they had stopped in front of an orphanage.

> You can imagine the sense of wonder on the kids' faces when all these westerners arrived handing out baseball caps and glasses. Then it all went dark! This made what was already a spectacular and rare event all the more memorable. Unbeknown to us, the conference

organisers were meanwhile panicking, thinking we had crossed into Cambodia and been captured by the Khmer Rouge, Nobel Laureate Ramsey and all.

* * * * *

By the time that 1999 approached, I was on the executive of the British Science Association, and found myself an adviser for a national committee planning for the solar eclipse. The only place on the British mainland where totality would occur was across Cornwall and the western extremity of Devon. At the committee's first get-together I met members of the Cornwall and Devon tourist boards, local government officials, and representatives of the police and emergency services. They wanted to know what to prepare for, how many people were likely to come, and what prior experience was there to draw on? At that stage, no one had much idea.

John Mason, an enthusiastic astronomer, and a veteran of several eclipses, was invited to describe the event. He bounded to the front of the room, next to the video screen and with infectious excitement exclaimed: 'A total eclipse is indescribable!' I was reminded of Mr Laxton, my schoolteacher from 1954, as John endorsed that promise from long ago: 'When you've seen a total eclipse for yourself, you will understand why it defies adequate description.'

Only those who have been present at the birth of a child can truly understand the emotional power of the event. The appearance of a new intelligence from the dark womb of its mother, the moment when the cause of that bump in the abdomen reveals its true identity, is overwhelming. If you have been there yourself, you will understand and may by proxy have an idea of the intense emotions that a total solar eclipse can engender. If not, then you will have to trust me, as I was at that stage having to trust John Mason.

He also explained the rarity of total eclipses. Although the last time one passed through Cornwall had been in 1724, he told us that in 1927 there had been a brief total eclipse, elsewhere on the British mainland. This was seven decades before 1999, admittedly, but newspaper accounts from the time survived, and gave some idea of what to expect. And there were many children from that time who were still alive. At our next meeting, we learned what happened in 1927, and from this began to make plans for 1999.

* * * * *

On 29 June 1927, the track of the eclipse had entered the mainland at Cardigan Bay in Wales, crossed Lancashire and Yorkshire, and exited near Hartlepool on the east coast of England (Figure 3A). The eclipse on that occasion had lasted a mere 24 seconds, and occurred soon after dawn, but nonetheless thousands of people had travelled large distances by train to see it. In addition, countless numbers within reach of the eclipse track had walked, cycled, or taken buses to experience this 'once in a lifetime event'.

Thomas Cook had advertised a special train, which would leave London's Euston station at 'about 11 a.m.' on the day before the eclipse and head for Southport in Lancashire. The excursion,

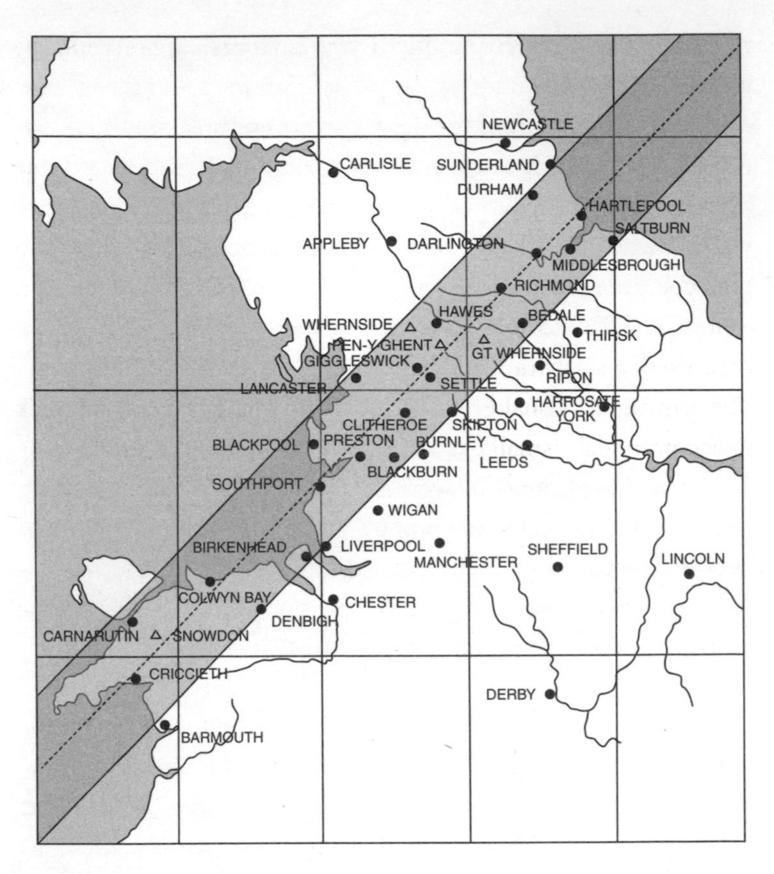

Figure 3A. Path of 1927 eclipse.

which cost £3 for 'third-class rail travel and a first-class hotel', would return to London the next morning, immediately after the eclipse had ended. This train belonged to the London Midland and Scottish railway. Not to be outdone, the LNER—London and North Eastern Railway—announced its own excursion from London's Kings Cross station, departure 10 p.m. on the night before the eclipse, due to arrive at Richmond in Yorkshire at 4 a.m. the next morning.

This left little margin for delays or, worse, breakdown, as the eclipse would reach totality just two and a half hours after the planned arrival. The return would depart at 2 p.m. As the journey would be through the night, there was no need for a hotel, and the cost came to a mere 18 shillings (90 p in today's decimal currency). A private company of entrepreneurs, Dean and Dawson, organised special trains for schoolchildren to go to Richmond. The army Educational Corps set up camps along the eclipse track to accommodate 'small parties'.

Weather conditions can potentially make or break an eclipse. This is true anywhere, and especially in England's unpredictable climate. Weighing the odds of where to find the best chance of success was a popular topic of discussion in the run up to the eclipse. Hartlepool, on the east coast, at first sight could have appeared promising, thanks to the potentially uninterrupted view of the newly risen sun across the sea. However, the Director of the Armagh Observatory advised: 'I know Hartlepool and its weather too well to want to go there.' He eliminated anywhere on the east coast due to the prevalence of 'dense, cold sea fogs in early summer'. The better the weather on the mainland, so the worse the sea fogs were likely to be, especially early in the morning. He concluded that midsummer at dawn in the higher elevations of the middle spine of England: 'gives the best prospect of clear sky that the British climate affords.' He favoured Richmond, Leyburn, or Giggleswick in Yorkshire.

For those still undecided, and with the means to travel, the BBC issued special eclipse forecasts in the two days leading to the eclipse. The Meteorological Office offered 'accurate short-range weather forecasts'; to obtain them you were instructed to 'call Holborn 3434, extension 174'.

* * * * *

We, in 1999, learned that our 1927 predecessors had encountered similar challenges to ourselves. As late as 30 March 1927, just three months before the actual event, the British Astronomical Association was discussing how to educate the public, not just on how to observe the eclipse safely, but also where best to go.

For example, many public posters showed the track but gave no warning that the duration of totality varies across its width such that you would need to be 'within seven or eight miles' of the centre line. Nowhere, apparently, had the million or so inhabitants of Liverpool been told that if they remained at home, they would have at best 1 or 2 seconds of totality. It was only after this became more widely understood that trains to Southport were organised to ferry Liverpudlians to the centre of the eclipse.

A 'Joint Permanent Eclipse Committee' was formed in 1927, which consisted of members of the Royal Society and the Royal Astronomical Society. The committee produced a map of the eclipse, scale ten miles to the inch, which showed the eclipse track, together with the duration and time of mid-eclipse throughout its length. A month before the eclipse, precise data led them to announce that 'the track should be moved 1/10 inch (one mile) in a north-west direction, parallel to itself'. The advice added the caution that as no one could be precisely certain about the outer edges of the totality track it would 'be wise to take up a position well within the zone of totality, in which case the correction will make no difference'.

* * * * *

As the 1927 eclipse had at most 24 seconds of totality, precise timing during the event would be necessary for anyone who planned to make serious observations. Above all, it is important not to be looking through an unfiltered telescope at the moment the sun reappears at the end of totality. Emotions are so high during an eclipse that all sense of time is easily lost. The best laid plans on how to use those precious minutes, or seconds in 1927, can go awry unless some metronomic plan is in place. The present day organised eclipse chaser may have a pre-recorded audio tape, which runs throughout the show, with timed instructions: 'take photo on f-stop this, bracket three around f-stop that, end of totality in five seconds, glasses on, diamond ring.' The result is that a photographic record is achieved, and the observer's eyes preserved, but all too often none of the eclipse is actually experienced other than through a camera's lens. Wiser enthusiasts include instructions: 'observe sky overhead with eyes for ten seconds, look around horizon, note the stars.'

The BBC's collaboration with the Royal Observatory in 1927 was the first chronologically automated eclipse. The sound of six pips was planned to be transmitted on the radio for each of 5 a.m., 5.15, 5.20, and 5.30 GMT, with single pips every second from 5.22 to 5.26, as the moon's shadow swept from west to east across the land. The BBC also promised: 'the minutes and every fifth second will be named verbally.'

Newspapers showed little interest in the upcoming event, until almost the last moment. On the Monday and Tuesday of eclipse week, however, a deluge of coverage filled their pages. The *Daily Express*, as if inspired by the account in Joshua, announced dramatically though erroneously: 'Sun to Stand Still on Wednesday.' In the last 48 hours, tens of thousands decided to head for the eclipse path.

The occasion was so extraordinary that three-quarters of a century later it remained fresh in the memories of octogenarians. Their testimonies illustrate the remarkable effect of a total eclipse in creating false memories as, in at least one case, the events described were not of totality.

Before dawn on 29 June 1927, Frank Symons' father woke him so they could share what he was assured was 'a once in a lifetime chance'. Without having had a bite to eat, they set out in the dark for their destination: a hill with a view across Lyme Bay. From this vantage point they could see all the way east to the horizon save for another hill, which was about half a mile away. Frank's dad evidently was not the only early bird intent on catching the lunar worm, as lots of people had gathered on the neighbouring knoll.

He remembered that before sunrise there was a light mist, which hovered over the bay and gave the proceedings a primeval quality. Their once in a lifetime experience was under threat, however, because the mist became thicker as dawn arrived. The sun was dimly visible as it rose, already just a crescent, low in the sky. The dawn had hardly happened before the moon was obscuring almost the whole of the sun. Twilight returned and the dawn chorus of birds stopped singing 'as if someone had thrown a switch'. Not just the birds, but everything went quiet. 'There was a deep and intense silence, quite eerie, and so complete that individual conversations between people on the other vantage point could be clearly heard,' Frank Symons told a reporter years later. Then it was all over, and as the sun rose out of the mist, the birds started to sing again and 'a disappointed 11 year old boy went home to his breakfast'.

Clear as Frank's memories were, they were not of a total eclipse as Lyme Bay in Dorset was far from the narrow track. Frank Symons had witnessed a partial eclipse, 95% of the sun obscured, and dramatic effects to be sure, but nothing compared to being underneath the night of total blackout. Frank would have had to travel over a hundred miles northwards to reach the edge of the full eclipse track.

Dora Ashden, who was 16 at the time, and lived near Liverpool, took one of the special trains before dawn and joined the crowds, en route to Ainsdale Beach, near Southport. The beach, which stretches for eight miles, fronts the Royal Birkdale golf course, and offers uninterrupted views across Liverpool Bay to the west. This is the direction from which the moon's shadow would appear. The sun, meanwhile, would be rising in the east, as the eclipse was due at 6.30 in the morning (05.30 GMT) of the British summer. The beach thus allowed a good panorama for the thousands who had flocked there.

Totality lasted less than half a minute, which is the length of time it took for the entirety of the moon's shadow to cross those beneath. A short time thus equates to a small extent of the moon's shadow, which in the case of the 1927 eclipse would have been only 30 miles in diameter. So to have any chance of seeing totality you had to be located well within that narrow band. Even at the very centre of the track totality would last only for 24 seconds, while towards its edges totality would be a mere second or two. Much of the conurbation of Liverpool, 20 miles to the south, lay outside the track, hence the special trains to take the crowds to Ainsdale Beach.

Unlike Frank, Dora certainly had the total experience. Dora recalled the eclipse as 'so memorable', the early morning gradually becoming darker until 'at the moment when the sun was completely obliterated there was a wonderful bright shining light all around; fiery all around the circle'. The moon's shadow had filled the sky above her, but with only half a minute of totality and the shadow so narrow, the dark arc filled only a small part of the sky and the surroundings were suffused with light. Joan Castleton, who had been 13 years old at the time, also remarked on this. She saw 'beams come out like a halo. It was wonderful—rays in all directions'. Even so 'it went very cold' and, like Frank Symons, Dora recalled: 'It was silent and really beautiful.'

Hilda Wooley lived in Blackburn, about 15 miles inland and eastwards along the eclipse track. She was ten years old, and with about 40 school friends, walked up Bunkers Hill on the southern outskirts of the town. She was scared by some of the older boys who said that 'the world comes to an end with the eclipse according to the scriptures, when the moon [shadow] catches the world'. She too recalled that it was 'dead quiet' as 'the birds stopped singing' and 'it went chilly'. Everything was 'pitch black' and 'very scary'.

* * * * *

In 1927 the population had a wide choice of venues, as the eclipse tracked across Wales and England from coast to coast. In 1999, however, on the mainland the Cornish peninsula would be the only possibility, and everyone with a car could potentially decide to turn up. To reach the far west, where the best chance of clear skies seemed likely to be, they would have to traverse the length of Cornwall across the moors by means of one or other of two major roads. Based on the evidence that in 1927 huge crowds flocked to see the eclipse at the crack of dawn, in an era when mass transport was limited and car ownership almost unknown, the authorities in Cornwall, 1999, began to plan for an invasion of over a million people in cars, buses, camper vans. A total solar eclipse is a personal event, but when magnified a few million times there was the potential for a disaster. Mild panic began.

How many beds are available in Cornwall at the height of the summer holidays? If a million people descend on the toe of England for 24 hours, how many portaloos will be required? Can hospitals cope with people who have been blinded by having their retinas fried?

For although you are unlikely to damage your eyes by looking at the sun normally—it is too painful and you look away long before real damage occurs—during a total eclipse there is a moment of danger when disaster can strike. This is at the end of totality, when the sun suddenly reappears from behind the moon. For 3 minutes, or however long totality lasts, your eyes have become used to darkness. Your pupils are enlarged. You will almost certainly be looking at the ghostly corona, the outer atmosphere of the sun, which is normally invisible but can be seen in all its glory during the darkness of a total eclipse. And you may be examining its tenuous strands by peering through binoculars, a telescope, or a camera lens, when—flash—the full intensity of the sun bursts through the valleys of the moon's silhouette. Billions of photons rush through your cornea at the speed of light, to be concentrated by the lens onto a blisteringly hot spot on the screen—your retina. One of the most important pieces of advice in the brochures, which were produced for the eclipse tourists, was that someone had to take responsibility to know exactly how long totality lasts, to time it, and to announce several seconds before it ends: 'Look away now!'

In case you head off to see a total eclipse before having read any more of this book, this may be an appropriate point to describe how best to observe the show. This was one of the contributions of John Mason to all at our meeting: spread the word on how to get the best experience of a once-in-a-lifetime event.

The show begins about 80 minutes before totality, when the moon cuts a notch from the outer rim of the sun. You won't be looking at this with the naked eye any more than you would normally view the bright sun, unprotected. Throughout all but the last moments before totality, the sun will be too bright to view directly, so how do you watch what's happening? You need some means to reduce the sun's glare, or to cast an image, which is bright but not blinding.

He told us that this danger had been known for a thousand years, and the solution went back several millennia. Al-Biruni, an Islamic scholar a thousand years ago, wrote: 'observations of solar eclipses in my youth have weakened my eyesight.' He recommended a trick, which had been used in antiquity by the Greeks, Egyptians, and Babylonians: look at an image of the sun on the surface of a bowl of water. The image reflects only a few percent of the total sunlight, but this is plenty enough to see a sharp sun without any discomfort. Any liquid will do, so if you are worried that a breeze will cause ripples, or that an over-excited eclipse chaser will bump into it in their rapture, use oil or some additive to make it less sloppy.

If instead you use a standard mirror, the scattered light will blind you almost as much as the direct sun. A solution is to use instead either a very small mirror, no bigger than a thumbnail in size, or, better, cover most of your large mirror with opaque material, into which you have already cut a hole no bigger than 1 cm, and then affix it to the surface of the mirror. You will, in effect, have made a small mirror, but in a form that is easy to handle. Orient the mirror so that it reflects the sun's image onto a wall or screen that is otherwise in shadow. This will give a clear, sharp, and large image of the luminous disc without any surrounding glare.

You will do best if the mirror is held fixed by some means, such as mounted on an easel, tripod, or a table. Ensure that it won't be obscured by shadow as the sun moves some 20° across the sky during the 80 minutes overture to totality. You can practise this any day without need for the added frisson of an eclipse. If you're lucky, you might even be able to see sunspots in the image. In the absence of special instruments, such as binoculars or telescopes with specially adapted sun-block filters, that is probably the best way to create an image of the sun.

If you plan to watch an eclipse with young children, this is an opportunity to expose them to some simple science: make a pinhole camera from a shoebox. If trees are in leaf their foliage illustrate the pinhole effect as gaps between their leaves cast a myriad of images of light on the ground. These are actually images of the sun—the gaps are like pinholes and the ground is the screen. On a normal sunny day it might not be immediately apparent that you are seeing images of the circular sun and not merely the illuminated gaps between shadows of the foliage. However, during a partial eclipse the bright array of crescent shapes can be remarkable. Any mesh will do. If there are no leafy trees, bring your own pinholes, as in the weave of a straw hat, or a colander. Images of the partially eclipsed sun within the shadow of the object make popular mementoes of the occasion.

Slightly more sophisticated, and ideal for a school project, is to make your own camera. You need a shoebox, some translucent paper, and a sharp point to make a pinhole. Cut off one end of the box and stick the paper sheet in its place; this will form the screen. All that's left to do is to make a pinhole in the opposite face of the box, and point it at the sun.

Rays that enter the box through the pinhole will create an (inverted) image of the sun on the translucent sheet. You will be able to watch the shape of the crescent evolve as the moon creeps across the solar disc.

Notwithstanding that these are simple ways to observe the sun's image safely, people will still want to have a peep at the 'real thing' with their own eyes. Specially designed glasses with foil in the eyepieces and cardboard frames to fit the goggles over your ears are easy to obtain. They are good so long as nothing unexpected happens. For example, you might be looking at the sun, safely, through your darkened eyepieces when there is a sudden puff of breeze, which dislodges them. Such wisps of cool wind can be common as totality approaches, and are all too likely to disturb flimsy cardboard spectacles. One moment you are looking at the obscured sun, and the next you are blinded as its full force hits your retina. So, the golden rule when using such goggles is this: keep one hand on their stem at all times.

Another test is to ensure that there are no scratches or holes to let sunlight through. A sheet of welder's glass is perfect, as it is large enough to hold in front of your face to view the sun, while also enabling you to turn your head and look around you as darkness falls. A good trick is to cut a rectangle in a large sheet of cardboard, and stick the welder's glass in that. An extreme solution is to put a whole box over your head, with the dark glass inserted in one side (Figure 3B). You might look strange, but at a total eclipse, no one will be paying you much attention.

If you adopt the more conservative approach of holding the glass in a single sheet of card, you can also make the container

Figure 3B. Safe viewing. Enclosing dark welder's glass in a cardboard box and wearing the ensemble keeps your hands free (photo: Frank Close). See Plate 5 for a colour version.

into a pinhole camera. Punch holes that spell out the name of your lover or child, let the sun cast a shadow of the card on the ground and within the shadow will appear their name spelled out in bright crescents. Who could fail to be impressed that, in the midst of nature's greatest spectacle, you were thinking of absent friends.

* * * * *

Thanks to John Mason's briefing, we now felt well prepared to enjoy the eclipse and to spread the word. For the south-west of England, here was an opportunity to attract tourists in record numbers. Ironic, and perhaps unique during its history, the tourist board actually tried to discourage visitors. It was afraid that the infrastructure would be unable to cope. If just one in a hundred inhabitants of Britain came along, of the order of a million would turn up. And with major cities like Bristol, Cardiff, and Birmingham little more than a hundred miles away, estimates of several millions could not be discounted. In short: no one had any idea of what to expect. The local government was seriously worried at the possibility that rural Cornwall could suddenly be filled with one million people. This is equivalent to the population of Birmingham, Britain's second largest conurbation, descending on the principality with the statistical certainty of deaths, injuries, accidents, breakdowns, food poisoning, lost children, assaults, and retinal blindness that this number would create.

Every threat is an opportunity. Enterprising farmers decided to hold back on planting seeds in some fields, and instead converted them to camper parks, complete with scores of portaloos. In the weeks before the eclipse, convoys of lorries laden with sanitary equipment made their lumbering way across the moors towards Land's End.

To ensure the best chance of a successful climax to my 45 years long pilgrimage, I had booked a camping spot at a school in Helston, where astronomers would be present. On 9 August, with the weather fine, and the forecast good, my wife, eldest daughter, and I joined the caravan of eclipse chasers on the M5 and headed west.

4

Cornwall: 11 August 1999

When you drive south through France along the Atlantic coast, you can tell when you cross the departmental border to Bordeaux because suddenly the hillsides are filled with vineyards. The soil has not abruptly changed at the boundary, such that vines prosper on one side rather than the other. Instead, it would seem that local farmers cannot ignore the opportunity to brand any wine with the appellation of Bordeaux, even if the truly great vintages, on which the reputation rests, are to be found several miles away, further south on the banks of the Dordogne and Garonne.

I was reminded of this discontinuity in August 1999 as I ventured westwards through the toe of England, from Devon into Cornwall.

Cornwall is older than England. Cornish folk are as proud of their heritage as are the Bordelais, and mark their borders prominently with both signs of welcome and proclamations of their independence. The boundary of the Duchy of Cornwall wanders to and fro as it follows the River Tamar, and other natural topography. For the eclipse, however, the northern limit of the path of totality would be a straight line. It would make landfall in north Cornwall near Port Isaac, and 90 seconds later leave at Teignmouth in Devon having toyed with their border near Launceston and Tavistock.

The presence of this invisible and ephemeral boundary could be inferred, however, by the change in farmland. No Bordelais transition to vineyards here. In its stead, countryside that in Devon and the eastern fringes of Cornwall had been devoted to agriculture and crops, now abruptly gave over to fields filled with portaloos, and the paraphernalia of campsites. We continued westwards for another 50 miles, diagonally through the path of the eclipse, across Bodmin Moor, much of which seemed to be prepared for an invasion by hordes of people who would need food, places to sleep, and sanitary facilities.

We bypassed the city of Truro, near the centre line. A climb to the top of its tower could have offered a good look out for the approaching shadow. We had another ten miles to go to our destination: Helston, near Penzance (see Figure 4A).

* * * * *

As August is in the middle of the summer vacation, a local school had rented out its vacant classrooms and playing fields as dormitories and campgrounds. A small hill near to the western boundary of the site gave a clear view towards the coast and the south-west. Land's End was 20 miles away in that direction, and the moon's shadow would reach it in two days' time at 11.10 a.m. and 19 seconds. Synchronise your watches! Thirty-eight seconds later it would envelop us in Helston.

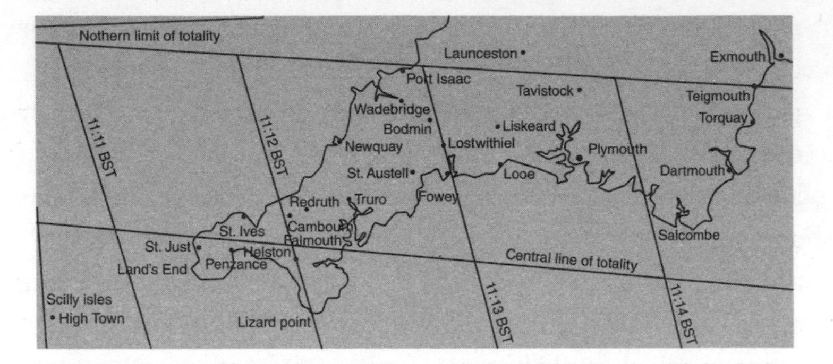

Figure 4A. The path of totality across South-West England on 11 August 1999, together with the times of maximum totality. The duration of totality straddles this by approximately 1 minute either side. Thus when totality ends in Helston, it has yet to start in Torquay (Source: http://www.nationalarchives.gov.uk/information-management/re-using-public-sector-information/copyright-and-re-use/crown-copyright/).

I prepared to experience my first total solar eclipse in the grounds of Helston School, 45 years after my initiation in the yard of my primary school in Peterborough. Helston in 1999 showed the progress in scholastic architecture in the interim. Whereas many of its buildings were the bland cubic and glass boxes beloved of the 1960s, its classrooms were light, open, friendly places to be. The grounds were extensive, with lots of grass.

It was on one of those grassy areas that we set up camp: ground sheets, two tents, and sleeping bags. The next day, 24 hours before the actual show, we joined other expectant campers in a dress rehearsal. An English August can be warm and humid, wet and stormy, or all of these in a single day. For our dress rehearsal, the sun shone in a clear sky on a day of rare perfection.

Members of a local astronomical society set up telescopes. They prepared demonstrations of how to view the eclipse safely, and to maximum effect. They covered a mirror with paper to remove glare and then cut a circular hole in the paper, no bigger than a penny, which reflected the sun's image onto a screen. One of their telescopes projected the sun's bright disc onto a wall at the end of a short, shaded, corridor. At some 2 m in diameter, the image was almost more dramatic than the sun itself. To our astonishment, and delight, it revealed sunspots. During the partial phases of the eclipse, these demonstrations would present images of the crescent sun, which could be viewed in complete safety, and to more dramatic effect than by looking directly with our eyes through darkened glass.

This was like my experiences from 1954, but in a grander form. At about 11 o'clock on 10 August we checked the position of the sun to ensure the instruments were correctly aligned. At that moment the moon must have been some 13° to the west of the sun, about the width of two outstretched hands, but impossible to see in the brightness. I imagined how in just 24 more hours the moon would be in a direct line, covering the sun like the lid of detached sky in my childhood fairy tale. What would it be like in reality? Forty-five years after my awakening, I was about to find out.

That night, remote from streetlights, it was dark. Constellations of stars twinkled in a cloudless, moonless, sky, and the shimmer of the Milky Way became visible as our eyes adapted to the gloom. The moon was already below the horizon, along with the sun, the couple making their final choreographed manoeuvres towards perfect alignment on the morrow. Excited, I closed the tent flaps, and settled into my sleeping bag full of anticipation. The law of gravity ordained the conjunction; nothing could prevent what was about to happen.

That an eclipse must happen precisely at 11.11 on 11 August 1999 had been predicted decades ago. If only we could predict the weather with such confidence. Sadly, I had not allowed for the vagaries of the British climate. When I opened the tent flaps the next morning I was greeted by a camper's nightmare: rain pouring from a leaden sky. By 9 a.m. the rain stopped, but that was the limit of the good news.

I had waited for this promised moment of natural wonder for nearly half a century. Now that the day had arrived, I felt like a character in some black comedy. How can you experience a total eclipse when the sun is already completely obscured by layers of impenetrable dark clouds?

* * * * *

I was not the only person in Northern Europe who hoped to lose their eclipse virginity after 45 years of lunatic celibacy. Afterwards, several friends told me their own stories.

Seven years old in 1954, Peter Boldon had watched the eclipse that June from his school in South Shields near Newcastle. Two hundred miles further north than Peterborough, he was that much nearer to the line of totality than I, and experienced a partial eclipse of nearly 90% of the sun. Like me, he had joined his classmates in the playground where they stood in four rows like soldiers on parade, and watched the unfolding drama. Were schools throughout England on eclipse watch that day?

He too had been told that there would be a total solar eclipse in 1999, and had made a note to see it. We had first met as prospective university students in 1962, unaware of a mutual interest in a solar eclipse that still lay more than three decades in the future. By 1999, he had a home on the Cotentin Peninsula in Normandy, a few miles south of the path of totality.

After leaving the mainland of England, near Torquay, at 11.15 British Summer Time, the path of totality would traverse the English Channel: La Manche. Brittany Ferries, whose ships commute between Portsmouth and Cherbourg, planned to stop in mid-channel during the minutes surrounding totality so that both passengers and crew could have a clear view. This would also reduce the chance of them running into other craft, skippered by distracted crews. Of the Channel Islands, Jersey and Guernsey lay frustratingly just south of the track, but Alderney was in its path. It then would clip the northern tip of the Cotentin peninsula of Normandy before carrying onwards across the mainland through France, Germany, and Austria, continuing towards Turkey, and eventually, nearly 3 hours later, India.

Peter had gathered about 20 friends and planned to drive with them to the northern tip of the peninsula to experience the drama under what Meteo France promised would be broken cloud. In Normandy, which was on Central European Time, an hour ahead of the United Kingdom, totality would commence at about 12.15. My Magna Carta date had come full circle for him, placing the eclipse well into the French lunchbreak.

From midday until around two o'clock, French business traditionally stops. Restaurants, by contrast, now open and do their major trade. What would they have planned for the day of the lunchtime eclipse? Peter checked with Pages Jaunes, the French Yellow Pages, and identified a local hostel with a restaurant. To be well prepared, and ahead of any rush, he called the proprietor several weeks in advance and asked if they could take bookings that day for 20 people.

Telephone calls in a foreign language are ripe for misunderstandings. The owner of the hostel, which had a large restaurant but only one bedroom, mistakenly thought Peter wanted 20 people to stay overnight. He burst into laughter as he replied: 'impossible, monsieur', in a way that Peter interpreted as proof that Normandy was already saturated with bookings, and about to be overrun with eclipse chasers. Peter now began his own panic wondering about these crowds, as rumours were spreading that the gendarmerie would close the peninsula to visitors, and he had already planned to drive to its northern extreme by back roads. He asked the owner how many people he was expecting, and only then discovered that the hotelier mistakenly thought Peter wanted a score of beds, rather than a table for 20. Lunch for 20, monsieur? 'Bien sur!'

* * * * *

In Germany, Rolf Landua planned to take his children to see the eclipse. A scientist, based at CERN in Geneva, Rolf's strategy was to visit his mother in Wiesbaden in the days before the eclipse, and then drive back to Geneva on the day itself. He calculated that if he set out at nine o'clock in the morning, he should enter the region of the total eclipse by about 11.00, and be around 200 km south of Wiesbaden, in the vicinity of Karlsruhe, and well within the eclipse path. With the total eclipse expected at 12.30, this would leave ample time for him to find a good spot where he and his children could watch.

He set off early in the morning to allow plenty of time to reach his destination, but it soon became clear that the majority of people in southern Germany also had the same idea. Traffic on the autobahn began to bunch, as waves of compression brought cars to a halt, after which they would move on for one or two kilometres until reduced to a crawl and then to a halt once more. This was repeated over and over, except that with each release from a bunch, the distance to the rear of the next blockage became less.

By 11.00, with totality just 90 minutes away, and with at least 20 miles still to travel, Rolf started to feel uneasy about reaching his goal. The autobahn resembled more and more a parking lot than a through way. Traffic was still moving, but slower and slower. His children started to grumble.

By 12.00, Rolf was almost at the limit of the total eclipse region, but now dense clouds were starting to obscure the sky. Adding to the misfortunes that seemed to be conspiring against him, a new phenomenon started to appear: many drivers decided that the autobahn could indeed also be used to park their car. Starting with the right lane, more and more cars had stopped, with their engines turned off; their passengers having got out were walking around leisurely. Driving became like slalom through an increasingly dense cloud of stationary cars and human hazards. Finally, Rolf decided to take the same approach: he parked his car and joined the crowd. Fortunately, he had just reached the extreme limit of totality.

Standing next to his car and waiting for darkness to come, Rolf and his children took their carefully prepared, cheap glasses (they 'only' cost 5 euros each) and stared into the sky, hoping for a gap in the clouds. And—indeed—just when totality was almost over, they could see a glimpse of the total eclipse for about 30 seconds.

* * * * *

Also in Germany, another scientific friend of mine, Hans Schmidt, had a similar experience to Rolf Landua. His plan was to take time out from a vacation in Bavaria and drive about 100 miles on the autobahn towards Munich, which was in the path of the eclipse. He too set off early in the morning but by ten o'clock, with totality just over 2 hours away, and with at least 50 miles still to travel, he was in gridlock. Cars overheated, children grumbled, families took emergency toilet breaks at the roadside, as people began to fear that they would not reach the path of totality in time.

Unlike Rolf, however, Hans decided that he would try to leave the autobahn. His new plan would be to travel anywhere in the general direction of where the moon's shadow was predicted to pass. He estimated that the eclipse track was roughly parallel to the autobahn, its centre point about 50 miles away. As the width of the track where totality would last for more than a blink of an eye extended for several miles, Hans estimated that he could probably get within the range of totality after driving another ten miles, or thereabouts.

He could see a road sign, ahead on the autobahn, which advertised an exit in a kilometre. A kilometre: you could walk it in less than 10 minutes, but it seemed to Hans that it took that long to reach it by car, such was the congestion. When he got there, at last, he discovered that several other cars were escaping the mayhem onto the exit ramp already. But at the top of the slip road they all selected to drive onto a major road, whose general direction mimicked that of the autobahn, and was itself also heavy with traffic. Hans selected a country road, which headed west, towards the incoming moon-shadow, and drove as fast as the narrow lanes allowed.

The skies were clear and the sun already in its final stages of partial eclipse. Those were the days before drivers relied on satellite navigation systems. Hans checked the map, drove for another 10 minutes, and then noticed a couple of cars at the roadside, with their occupants gazing at the sky. Trusting that he must now be within the path of totality, he and his family joined them. It was midday and the sky was clear; totality was less than 20 minutes away.

* * * * *

Meanwhile, in England, I was carefully placed on the centre line of the eclipse, but under thick cloud. I feared that the eclipse would come, and go, and that I would see none of it.

According to experts, as the moon's shadow approaches, the air cools and clouds can disperse. While that might well happen for light cloud cover, I doubted that there would be sufficient time for the banks of dark cloud in Helston to clear. Nonetheless, I continued to hope for a miracle.

By 11.00, with just 11 minutes to go before totality arrived, the gloomy day was, if anything, even danker. At least it was obvious that something dramatic was happening. The last glimmers of daylight were being rapidly obliterated, but I had by now resigned myself to frustration: the show of a lifetime was happening but on the far side of the curtain, out of sight. By the time the storm had cleared, the eclipse would have moved on to Europe and beyond, never to be seen again.

I had waited most of my life for this moment. Since that 1954 June day in Mr Laxton's class, a total of 541 months had elapsed. At 11.02, there would be just 541 seconds more to wait. By 11.09, and with now less than a mere 100 seconds before show time, it was obvious to even the biggest optimist that we would not see the eclipsed sun. All I could do was to look to the west, from where the shadow would come, in the hope of finding some small break in the clouds.

But there was none. Indeed, if anything the view was even worse. In the far distance, blackness signalled what appeared to be a storm of gargantuan intensity gathering just beyond the western horizon, near Land's End. Then I realised: this was not further bad weather, but instead was the dramatic overture to the main event.

The vision was terrible, as if I was present at the Apocalypse. Like a giant tornado filling and gaining strength, the blackness swelled out sideways and saturated the distant sky in less time than it takes to articulate. This lateral motion was the result of a tsunami of darkness rushing towards me, swallowing up the intervening space at the rate of a kilometre every second. Within half a minute the moon's shadow had enveloped the entire landscape, in utter silence. It was as if a black cloak had been cast over everything.

With no chance of seeing the eclipse directly, and thanks to the lack of any other diversion, I had chanced upon a unique experience of the moon's shadow, something that I have never again witnessed in eclipses under clear skies. The clouds were already dark,

Plate 1. Photo of a total eclipse (photo: Frank Close). See Figure on page vii.

Plate 2. Total eclipse over the Sahara. A meteorological satellite's view of the moon's shadow as it crossed the Sahara Desert during the solar eclipse of 29 March 2006. Had this been a high-resolution image from a spy satellite, the author would be visible near the centre of the silhouette. The full story is in Chapter 7. A moving image of the shadow is at http://bruxy.regnet.cz/web/meteosat/EN/solar-eclipse-2006/ (Credit Line: © 2006 EUMETSAT, Meteosat 8 image received and processed by Martin Bruchanov.) See Figure on page 2.

Plate 3. Shadows and crescents. The lattice of a chair creates a matrix of pinhole cameras, which project crescent images of a partially eclipsed sun onto the floor (photo: Frank Close). See Figure 1A on page 9.

Plate 4. Orange moon. The appearance of an orange moon during a lunar eclipse. (Photo courtesy of NASA.) See Figure 2A on page 26.

Plate 5. Safe viewing. Enclosing dark welder's glass in a cardboard box and wearing the ensemble keeps your hands free (photo: Frank Close). See Figure 3B on page 69.

Plate 6. Eclipse history on a floppy hat (photo: Frank Close). See Figure 5B on page 91.

Plate 7. The corona during the 1878 eclipse. A hand drawing of the corona by Etienne Trouvelot, as observed during the eclipse of 29 July 1878. (Source: https://commons.wikimedia.org/wiki/File:Trouvelot_-_Total_eclipse_of_the_sun_-_1878.jpg) Total eclipse of the sun. Observed July 29, 1878, at Creston, Wyoming Territory. (Plate III from E. L. Trouvelot, *The Trouvelot Astronomical Drawings*. New York: C. Scribner's Sons, 1881–1882). See Figure 5D on page 101.

Plate 8. Eclipse goggles in Zambia (photo: Frank Close). See Figure 6A on page 117.

Plate 9. Saharan traffic jam: rush hour en route to the path of totality in the Sahara. Expect traffic jams in the USA 2017 to be even more extreme (photo: Frank Close). See Figure 7A on page 132.

Plate 10. 'Solar eclipse watching point 500 m' (photo: Frank Close). See Figure 7B on page 141.

Plate 11. Preparing for totality (photo: Frank Close). See Figure 7C on page 142.

Plate 12. 'The crow and the crescent', an 80% partial eclipse in Abingdon-on-Thames, 2015. This eclipse was total in the vicinity of the Faroe Islands (photo: Frank Close). See Figure 9B on page 195.

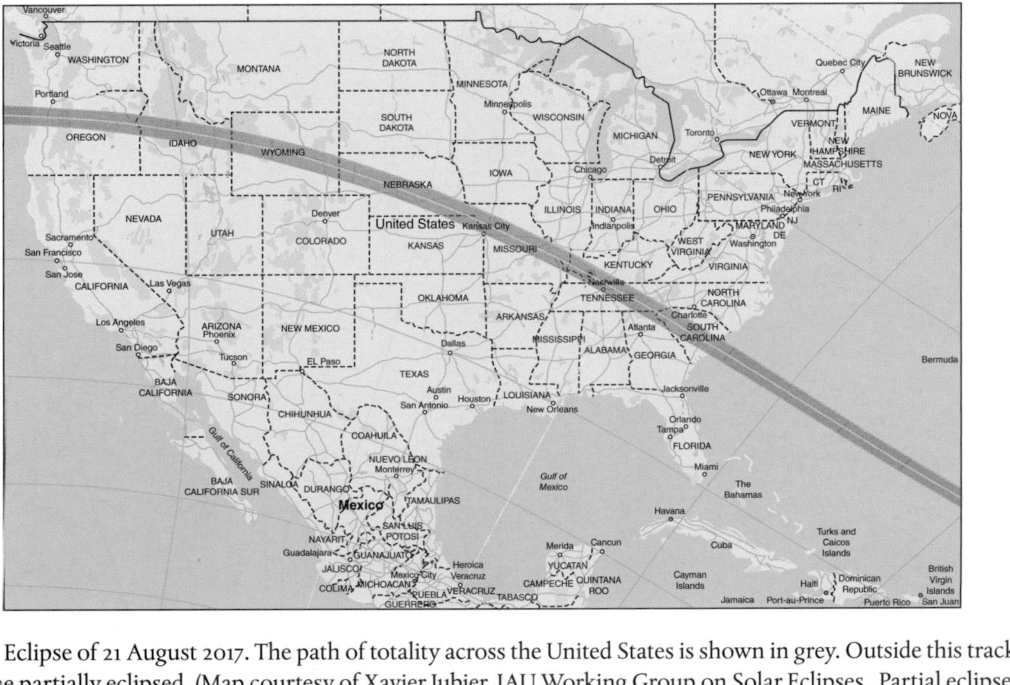

Plate 13. Eclipse of 21 August 2017. The path of totality across the United States is shown in grey. Outside this track, the sun will be partially eclipsed. (Map courtesy of Xavier Jubier, IAU Working Group on Solar Eclipses. Partial eclipse data after Mark Littmann, Fred Espenak, & Ken Willcox, Totality: Eclipses of the Sun (Oxford: Oxford University Press, 2008), 'The path of totality visits substantial parts of 11 states,' p. 256.) See Figure 10B on page 201.

but darkness was hidden by darkness, as the coal-black shadow from the obliterated sun splayed on the overcast. If every cloud has a silver lining, then every clouded eclipse has a deep wall of darkness as its herald.

This awesome experience, where I felt insignificant in the presence of a unique natural phenomenon, made me even more anxious to see the cause of it all. Cornwall's once in four centuries show, today, for 3 minutes only, was taking place right then, directly above my head, but out of sight. If only the clouds would evaporate. I watched, frustrated, as if seeing a silent film projected onto a cinema screen from behind.

Precious seconds passed as I willed the clouds to part. 'If mind can influence matter, please let it do so now', I prayed, in case any director was listening who could affect the choreography. The clouds remained, however, like an opaque curtain separating audience from the stage. Did no one realise that the show had begun and would soon be over? Perhaps a more direct challenge was needed. 'If there is a God who could stop the sun and moon for Joshua, and who continued to ordain miracles on a regular basis 2000 years ago, now would be an ideal moment to make a convert.'

If what happened next was anything other than coincidence, then either heaven is very near to Cornwall, or communications with it travel faster than light.

In the distance it looked as if the clouds might have parted a little. About a mile away, cameras flashed like at an Olympics medal ceremony where the crowd want their personal record. At least in the Olympics there is the chance that a camera might benefit from the added light. To hope to illuminate the rear of the moon, which is some 250,000 miles away, is silly. A golden rule at any total eclipse is to block your flash. Not only does it achieve nothing for your photo, it is an annoyance to everyone as light is the last thing anyone wants once eyes have become accustomed to the gloom of the eclipse.

There could have been only a few seconds of totality that remained, before daylight returned and the shadow moved on to engulf Alderney, Normandy, and the mainland of Europe. All around us now, the horizon was relatively bright. As if to complete the irony, only in these last seconds of totality did the clouds thin. My plea had been heard, or perhaps the folk wisdom was true after all—atmospheric changes are so dramatic that a microclimate tracks across the globe in and around the blackout.

Then a voice exclaimed: 'Is that the sun?'

I looked upwards. I noticed something undefinable, for the briefest of moments before clouds enveloped the mirage once again. No bigger than a fingernail, it looked like a dark spot on a silver cloth, nothing more.

Presumably that was the eclipsed sun, as I could find no other explanation that matched what I had just seen, and it was in the right part of the sky. Then as quickly as the clouds had parted, they closed again, but the eclipsed sun—I had seen it! Yet it was so small that I could not relate it to the vastness of the shadow that had filled the entire landscape. The canopy of darkness projected onto the clouds had overwhelmed me. I imagined this must be like the experience of a tiny insect in its final moments as a bird of prey, wings outstretched, swoops to devour it. That pimple in the sky, like an insignificant black dahlia with petals of silver gossamer, is deceptive. In reality the lunar surface is as large as the whole of

Africa. The moon's shadow had crossed about 250,000 miles of space. When it reached the earth, the line of sight obscured the sun for over 10,000 square miles, and it was this vast expanse that the cloud screen had intercepted.

I felt insignificant. Overwhelmed. Humble in the presence of a great natural event. I knew then, if not already, that I had to witness a total solar eclipse again, if only to be sure that that weird sight was indeed what I thought it to be.

Now we had one final task: to beat the rush on the roads out of Cornwall. This was when we found that there was none. The warnings of potential disaster had been so effective that when 11 August finally arrived, many stayed away. Fields, empty of crops, were devoid of campers too. Portaloos remained in pristine condition. The authorities' attempt to discourage an invasion by millions of visitors had worked.

5

June 21st is Midwinter's Day

Four days later, on Sunday 15 August, two dozen people who had been in Cornwall for the eclipse, but had never met, were reading the weekend newspapers when they had the same idea. It was this: 'Total eclipses of the sun often happen in exotic places, so trips to see them might take us to some interesting locations.' Which inspired the thought: where's the next one, and how long do we have to wait?

In 1954, I would have had to ask Mr Laxton, and hope that he knew the answer. In 1999, however, thanks to Tim Berners-Lee (born 1955) we had the World Wide Web. I entered 'total solar eclipse' into an Internet search engine, and back came the answer: 21 June 2001.

The wonders of Berners-Lee's invention, and a few clicks on my computer keyboard, told me everything about the eclipse. The narrow path of totality would begin at sunrise, just off the coast of Uruguay, then sweep across the South Atlantic, and make landfall on the African continent in Angola. It would transit Zambia, Zimbabwe, and Mozambique, and then depart the east coast to cross the Indian Ocean. Finally, it would clip the southern tip of Madagascar at sunset. A partial eclipse would be visible over much of the southern hemisphere, encompassing South America, the South Atlantic, and Africa, south of the Sahara. Thus it would be on their midwinter's day.

The 19 years cycle of eclipses showed that there had been a partial eclipse on the same day, 21 June, in 1982. That one had crossed the southern oceans off the coast of Antarctica, its extent limited by the darkness of midwinter. Looking to the future, I found a subsequent eclipse to be due on 21 June 2020, which would again cross Africa, India, and China, but annular. These differences arise because the moon oscillates back and forth, from apogee to perigee, as if on a spring controlled by gravity. Thus although the nodes recur on the same day of the year, the moon's distance from the earth alters. It would be relatively remote in 2020, and hence not fully obscure the sun, whereas it would be near enough in 2001 to make a total eclipse. As I had no reason to wait another 19 years, the continent of Africa seemed to have become my vacation destination for 2001.

Now I had to turn the dream into reality.

Not everything had yet been trapped by Berners-Lee's Web, as I found the perfect eclipse expedition advertised in the travel supplements of the newspapers. An enterprising tour operator offered flights between London Gatwick and Lusaka, a safari in the Zambian bush, and the total eclipse as a bonus. I signed up immediately.

I learned later that the trip was sold out within days. If the first rule for eclipse chasers is to look away at the end of totality, here is the second: those who hesitate are lost.

It can be expensive to be an eclipse chaser, but with 18 months on average between them, there is time to save up, be selective on expenditure in the interim, and to make preparations. The plan was first to fly to Lusaka in Zambia, ten days before the eclipse. We would spend a week at a bush camp in the Kafue National Park, 200 miles north-west of Lusaka, and then fly on a light aircraft to the grandly named Royal Airstrip in the Lower Zambezi National Park. We would be met there, and then driven in jeeps to a bush-camp on the banks of the Zambezi, for the eclipse itself. The decision made, the deposits paid, all that remained was to wait for 22 months.

But first there would be the Millennium, and the infamous bug. There was a widespread fear that on 1 January 2000, as computer accounts dated 99 reverted to 00 instead of moving upwards to the value 100, there would be a global systems collapse. This oversight in computer operations was known as the Millennium Bug. As it turned out, nothing untoward happened. Our computerised reservations for travel to Zambia remained safe, as did evidence that we had paid our deposits. The letter assuring us that all was well also included demands for the final payment.

* * * * *

At last the day of departure arrived.

We eyed the passengers on the train from Waterloo to Gatwick. They seemed a normal mix of travellers, set for business or pleasure, near or far, but there was no evidence of anyone prepared for the twilight zone of a solar eclipse. There were no bags, such as ours, whose labels ostentatiously sported the logo of the eclipsed sun. We wondered who our companions would be, and how many others were at Gatwick for a trip to the dark side.

There were probably more than a thousand travellers in the departure hall, but you don't need to be a trained detective to deduce the goals for many. Sun-seekers were already dressed in beach clothes and floppy hats; there were sports fans in their team colours and hikers wearing solid boots at the height of summer. There were no obvious signs of any eclipse chasers in the throng.

The eclipse would cut a narrow track across Zambia. The actual locations where the chasers were headed varied widely, but we all had to get to Lusaka first. And when we joined the meandering snake of travellers at the Lusaka check-in desk, we found that the whole flight was filled with lunatics. It was as if we had entered a parallel universe. For eclipse chasers have a uniform, like train-spotters. Cameras and binoculars were slung over their shoulders, telescopes and tripods corralled under their arms, manuals and almanacs poked from their pockets. They wore floppy hats (Figure 5B), anoraks, and even T-shirts that proclaimed their eclipse histories. One sported a shirt emblazoned with the headline 'Totality World Tour', atop a list of eclipse dates and venues, which made the solar eclipse show appear like some heavenly rock band on the road.

For the first time we were aware of the eclipse-chaser phenomenon. There are memento buttons, like campaign medals, which you can collect and display by pinning them to your shirt or hat. It is taken for granted that you were present at the eclipse in 1999, 2001, or whenever; bragging rights are determined by where you witnessed the phenomenon. Cornwall as a location in 1999 qualifies, certainly, but pales—literally and metaphorically—relative

Figure 5A. A Totality World Tour t-shirt (photo: Frank Close).

to the Black Sea coast of Turkey, Iraq, the mountains in Iran, Pakistan, or the borders of the Arabian Sea. Any of those locations would have given me a better view of the eclipse than I achieved in Cornwall, and the celebratory T-shirt would have impressed more people when I displayed it in Oxford. For solar eclipses have spawned an industry in the manufacture of memorabilia, which

Figure 5B. Eclipse history on a floppy hat (photo: Frank Close). See Plate 6 for a colour version.

display the path, the location—latitude and longitude in degrees, minutes, seconds, and fraction of a second—and some artistic rendition of totality or the scenery to complete the ensemble.

I discovered that there is also a private language in the fraternity, which was like a secret society, with its rituals and mantras. The cramped waiting lounge buzzed with excited chatter. 'What's the zenith?' 'It's saros 127.' 'What f-stop is best when photographing the diamond ring?' 'I've brought a filter for the hydrogen alpha line.' And more besides.

It was soon apparent that almost everyone in the lounge was flying to Zambia with the same intent. From their chatter I began to learn about not just eclipses, but also about the characters that make expeditions to the far reaches of the earth to see them.

Some really intense eclipse chasers keep a record not only of the number of total eclipses that they have seen, but also the accumulated time spent beneath the moon's shadow. This made me nervous about what I was taking on. My goal was to see a total eclipse, even one eclipse, so long as it was under a clear sky. I had no plans to 'collect' eclipses, even less to sum my total time in darkness.

While I could understand keeping account of the number of eclipses—I, after all, was very much aware that for me this would be number two—the reason why one would want to keep track of the total amount of time in the dark, left me in the dark. Suppose that you have travelled by plane, bus, and camel to the heart of Mongolia to witness an eclipse of just 30 seconds. Its brevity in time reflects the narrowness of its path, so to be sure of being able to see it you must determine your location carefully to a precision of a few hundreds of metres. If you make a mistake, the duration of totality will be shortened, or even vanish completely. So in my opinion, successful observation of a short duration eclipse is a greater adventure, and should carry more kudos, than to be present for several minutes of darkness. However, if you want to experience the wonder rather than collect eclipses like a botanist or train-spotter, longer is better.

Having made that argument to one of the throng, I unwittingly became a member of the club. One of the enthusiasts, an animated man whom central casting would choose as their computer geek, duly informed me that this was a widely held opinion. He explained that for a real eclipse aficionado, the measure of lifetime quality is the number of eclipses, squared, divided by the total time in minutes. This obscure formula has some basis, he added, and began to explain why.

By now, it will be apparent that I had already lost my soul and joined the cult. I had signed up for an eclipse adventure, and paid a small fortune for the experience. I had an hour to wait until the flight was called, so I decided that I might as well hear what he had to say. I listened as he explained that the formula gives credit for the number of times, weighted by their average duration, where shorter is better. 'If you measure the time in minutes and multiply the final answer by three and a half, the average target score is one', he enthused. 'Why is that?' I asked, partly because I was now hooked, but also because I suspected that someday I would write about this strange encounter and readers would want to know. He explained that three and a half is approximately the median time in minutes of a total eclipse, so if you put one eclipse for three and a half minutes into the formula, it would be par for the course.

To show how quickly I had become assimilated into the sect, I was able to converse with other chasers in the waiting lounge and proudly announce that I scored an impressive one-point-one. Based on but a single eclipse, which had been totally under cloud, this struck me as somewhat disingenuous, but it was an opening gambit, like a verbal form of Masonic handshake, which accepted me as one of them. We had not even left England and already, in the Gatwick departure lounge, I was comparing meaningless statistics with complete strangers.

* * * * *

Although everyone that I had spoken to up to that point was an eclipse enthusiast, I discovered that not every passenger on the aircraft was an amateur on vacation. Among the 200 travellers were a handful of professionals, en route to Lusaka to make experimental measurements of the sun's corona during totality.

I was intrigued. I knew that historically it had been during total eclipses when astronomers observed the sun's corona. This ephemeral regime of hot gases extends far beyond the sun's visible surface, but is normally swamped by the intensity of the sunlight, and so is lost from view. During totality, however, normal sunlight is dimmed, and the corona can be seen in its full glory. Electrically charged particles whirling in the hot ionised gases of the solar atmosphere form wispy tendrils in the corona. These reveal the magnetic fields surrounding the sun, much as iron filings reveal the fields surrounding a magnet in a laboratory demonstration.

Whereas in the distant past astronomers who wished to observe the corona had to wait for a natural eclipse to occur, then travel the globe to meet it, and hope for good weather, modern scientists have technology to help them. Satellites can carry telescopes up into space. High above the stratosphere where there is no air to scatter the sun's rays, the powerhouse of the solar system shines as a clear sharp white disc in a black void.

The coronal tendrils are like strands of gossamer, which extend out to a distance of several solar diameters. Specially designed shades on the satellite's cameras can blot out this bright sun, and display the faint corona in what is in effect an artificial eclipse. This is fine if your goal is to study the corona far beyond the solar disc, but to access the corona's innermost regions, you must wait for a natural total eclipse, where the moon obscures the sun. This is the main scientific interest of total solar eclipses today.

Still left with time to kill before we would board our flight, and by now saturated with geek-speak, I welcomed the chance to pick a professional's brains. In addition to having gained a glimpse of why total eclipses remain interesting today, I received a potted history of how eclipses had been used for the purposes of science, even back to ancient times.

Among the philosophers of ancient Greece, Pythagoras is famous for his theorem about right-angled triangles. Less well known is that around 500 BC he seems to have been the first to propose that the earth is a sphere. There was no experimental basis for this hypothesis, however. Pythagoras reasoned that as in three dimensions the sphere is the perfect shape, the Gods would have created the earth with this perfect form. About 150 years later, around 350 BC, Aristotle used a lunar eclipse to demonstrate that indeed the earth is round.

Aristotle was aware that sunlight casts shadows on the ground, whose shapes show the outline of whatever obstructs the sunbeams. He realised that at night we are in the earth's shadow. Then he reasoned that if the moon passed through the shadow, an outline of the earth could appear on it. Thus, at a lunar eclipse, the shape of the earth's shadow corresponds to the profile of our planet.

His insight was metaphorically as brilliant as the sunlight that the earth had interrupted. Whereas today this might appear to be obvious, for Aristotle it was a remarkable abstraction. First, he had to be aware that sunbeams travel in straight lines, and maintain this all the way to the distant moon. Second, the moon is itself a real solid body, capable of acting as a screen on which an image can be projected. Third, that the curved dark arc across the moon's face is indeed the shadow of the earth and bears no relation to the darkness of the familiar crescent moon, which waxes and wanes during its monthly cycle as a result of the changing angle between earth, moon, and sun.

That the earth is curved was already apparent, as when a ship disappears over the horizon, or when the surface is viewed from a high mountain. Aristotle had earlier noticed that as he travelled south, stars came into view that had been previously hidden below the southern horizon. Nonetheless, philosophers at that time argued vigorously about the shape of the earth. Thus we have the fourth, and in my opinion the most significant aspect of Aristotle's insight: this was an empirical test. Long before the scientific revolution and the Enlightenment of the seventeenth century, Aristotle's proposal had used the experimental method to settle a philosophical debate.

Presumably the Greeks made Aristotle's test during a partial eclipse, as it is then that the earth's profile is visible. At a total eclipse, by contrast, our atmosphere bends sunlight back onto the moon's surface. This leads to beautiful effects, as I myself saw in California, but there is no obvious hint of the earth's shape in this.

The use of a natural astronomical phenomenon to exhibit the earth's curvature, then led to quantitative measurements of its size. Convinced from lunar eclipses that the earth is spherical, in 240 BC Eratosthenes used the sun, and geometry, to measure its curvature, and hence its size (Figure 5C). He lived in what is now Egypt, and at the noon of the summer solstice in southern Egypt, the sun is directly overhead. He knew this, because there was one day a year—the solstice—when the sun cast no shadow at the

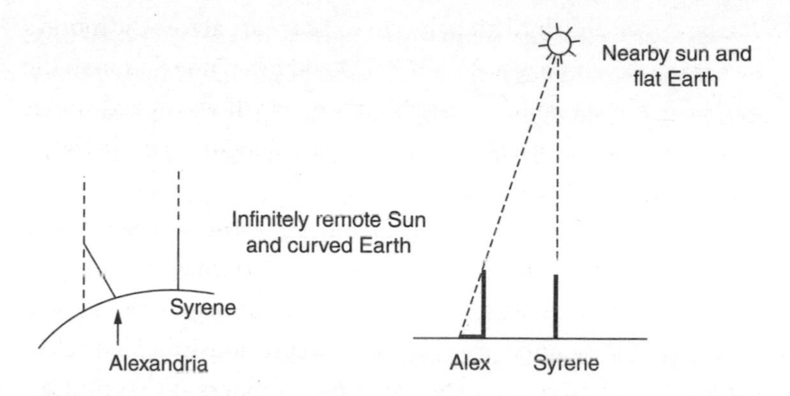

Figure 5C. How Eratosthenes measured the earth's circumference. The diagram on the left illustrates the shadow in Alexandria as due to an infinitely remote sun and a curved earth. The diagram on the right illustrates an alternative: that the earth is flat and the sun is nearby (photo: Frank Close).

bottom of a well in Syrene; in other words, it was directly overhead. Eratosthenes then argued that in Alexandria, further north, a vertical stick would point in a direction that differs from the 'true vertical', which was defined by the well-shaft further round the curve in Syrene. The angle between this stick and the well would be revealed from the length of the stick's shadow at high noon on the special day.

He did the experiment, and measured the angle of the shadow in Alexandria. It was about 7°. A complete circle contains 360°, so this told him that the distance from Alexandria to Syrene is a fraction 7/360 of a complete circle. In other words, the arc from Syrene to Alexandria is 7/360 of the circumference of the earth. Eratosthenes' answer for this distance came out at 25,700 miles, which is remarkably good: the true value for a circuit of the earth is about 24,900 miles.

This is also a moral tale in drawing inferences. Prior to Aristotle's demonstration that the earth was round, many had believed it to be flat. With that erroneous assumption as their foundation, Egyptian mathematicians had used the differing lengths of shadows in different locations to deduce that the sun is a few thousand kilometres away. After the observation of the earth's circular shadow cast on the moon, the assumption of a round earth revealed that the 'few thousand kilometres' is actually a measure of the curved earth, not the distant sun. Eratosthenes had implicitly assumed that the sun is much further away than the size of the earth, which is by chance true. The importance of the lunar eclipse was thus the ability to discriminate between two alternative hypotheses about the shape of the planet, and to eliminate the possibility that a sun nearby illuminates a flat earth.

* * * * *

While a partial lunar eclipse reveals the shape of the earth, so a total solar eclipse exhibits the outline of the moon. While that might appear trivially obvious, nonetheless the actual observations initially led to huge confusion, thanks to the discovery of what we now know to be the solar corona.

Following a total solar eclipse in 90 AD, Plutarch in Greece remarked: 'A kind of light is visible around the rim [of the moon] which keeps the shadow from being profound and absolute.' This is possibly the first mention of the corona, although Plutarch might have been describing an annular eclipse, where the moon does not completely fill the solar disc and leaves a bright ring, or annulus, visible.

The German mathematician and astronomer, Johann Kepler, seems to have been first to draw particular attention to the corona, following a total eclipse, visible in Prague, in 1605. He described the moon's 'blemishes' and asked: 'Why does the moon remain partly visible when the sun is behind it during total eclipse?'[1] His answer: 'The whole moon is surrounded by some sort of aery essence, which reflects rays from all parts.' He went on: 'There is some body around it which is translucent . . . less subtle than the rest, which, like a window, does not so much reflect the light of the sun as retain it.' As the sun, normally, appeared to be a bright circle, Kepler associated this wispy colouration with the moon. But what is it?

During eclipses, the observations of the corona began to take on special interest, which persists to this day. Astronomers in the seventeenth century took Kepler's discovery as evidence that the moon has an atmosphere, which is only visible at an eclipse when backlit by the sun. Yet when they looked at the silvery moon through telescopes, they found no evidence of any atmosphere. There were no clouds, nor any dust storms that would have shown the presence of wind.

Even as late as the eighteenth century, the corona was mistakenly thought to be the lunar atmosphere, notwithstanding its complicated structure, which is in general far from circular. Edmond Halley in 1715 attempted to explain its irregular shapes with the argument that the sun heats only one face of the moon at a time, which causes a large cloud of gas to develop over that side more than elsewhere. In Halley's opinion, this causes an asymmetric discharge, and hence the complex forms of the corona.

[1] J. Kepler, *Astronimae pars optica.*

Halley was wrong, but more than a century passed during which the corona was believed to be the lunar atmosphere. Not until the end of the nineteenth century was the correct explanation found.

Better telescopes, and observations over many eclipses, confirmed that the corona extends far beyond the lunar disc. Its shape changes from one eclipse to another. In 1871 it was almost circular, like the moon itself, whereas in 1878 it appeared to be elongated (Figure 5D). Astronomers noticed too that the corona appeared to have streamers, which resembled the magnetic lines of force revealed when iron filings surround a bar magnet. Furthermore, the changing shape of the corona seemed to be correlated with the varying cycle of sunspots—the bright sun exhibits occasional dark spots, which come and go over a period of about 11 years. Even though their origin would not be understood until the twentieth century, sunspots are clearly on the surface of the sun. So a link between the sunspot cycle and the shape of the corona confirmed that the corona too is solar and not part of the moon.

I have written 'confirmed' because the eclipses of 1868 and 1871 had already given the first clues to the solar nature of the corona. This came about thanks to a new scientific fascination: atomic spectra—the ability of different elements to shine with characteristic pallets of colour.

If an atom is hot, photons—particles of light—shake loose from the atom's electric fields. When this happens, photons emerge with energies that are characteristic of the parent atom. When our eyes absorb photons with these different energies, our brain interprets them as having a range of colours. Atoms do not only emit photons, they can also absorb them, but only if their colours

Figure 5D. The corona during the 1878 eclipse. A hand drawing of the corona by Etienne Trouvelot, as observed during the eclipse of 29 July 1878 at Creston, Wyoming Territory. (Plate III from E. L. Trouvelot, *The Trouvelot Astronomical Drawings*. New York: C. Scribner's Sons, 1881–1882). See Plate 7 for a colour version.

match those characteristic of the atomic elements. This led to discovery of the element helium in the corona.

The hot sun, like all stars, emits electromagnetic radiation across the entire spectrum. There is a lot of gas in its outer atmosphere, containing a smorgasbord of elements. In sunlight, those photons whose energies—colours—match those characteristic of the atoms of those elements are absorbed, and never reach the earth. These 'missing' photons show up as dark gaps in what, at first sight, appears to be a continuous spectrum of colours.

These lines, which were first seen in sunlight in 1802, are like some fundamental bar code, which identifies the elements present in the corona. They are known as Fraunhofer lines, after the German, Joseph von Fraunhofer, who made an intensive record of them during the following decades. It was another German physicist, Gustav Kirchhoff, who in 1859 broke the code and showed how the patterns reveal the presence of specific atomic elements. Following Kirchhoff's breakthrough, astronomers eagerly awaited a total solar eclipse, so that they might to learn about the nature of solar prominences—red flames, which had been seen to erupt from the rim of the sun.

On 18 August 1868, a total solar eclipse passed over India, the Malayan peninsula, and Thailand. A large number of astronomers gathered along the track. Among them was the Frenchman, Pierre Janssen, who set up his telescope at Guntur near the Bay of Bengal.

All observers that morning discovered that the spectra of the prominences revealed the unambiguous presence of hydrogen. And for most of the watching astronomers, that was that. Once the eclipse had ended, they repacked their equipment, logged their results, and prepared to return to their homes. Janssen, however, had an inspiration, and stayed on to take more measurements.

Janssen had noticed that the spectral lines had been so intense that it might be possible to study them without need of an eclipse. He had also noticed a yellow line in the data, which didn't match with any known element. That afternoon the sun was obscured by cloud, but the following morning was clear. Janssen pointed the slit of a spectroscope to that part of the sun's rim where, the previous day, the largest prominences had shown.

He succeeded in locating these subtle bulges on the solar perimeter. He then improved his apparatus, so that it could observe the rim of the sun, and with it he studied the spectra of prominences for 17 days. He confirmed the presence of a novel yellow line in the spectrum, and proposed that it must come from an element, hitherto unknown. Today this is known as helium—after Helios, which in Greek mythology was the personification of the sun.

By chance, two months later, an English astronomer, Joseph Lockyer, also saw the helium line in the spectra of prominences, in normal daylight. Lockyer knew nothing of Janssen's data and, by coincidence, both of their papers arrived at the French Academy of Sciences on the same day. Their independent claims to have discovered a new element, however, were received with scepticism and even ridicule. Eventually they were shown to be right, 30 years later in 1895, when helium was discovered on earth, as a gas encased within uranium rocks. (Today we know that the natural radioactivity of uranium emits the nuclei of helium atoms.)

During the total eclipse of 1871, Janssen saw Fraunhofer lines in the spectrum of the corona. These revealed the presence of sodium and hydrogen in the corona, added to the growing evidence that the corona is solar, scatters sunlight, and extends several solar diameters beyond the sun's bright rim.

Thus by the end of the nineteenth century, the corona was finally identified as being the remote atmosphere of the sun itself, and nothing to do with the moon at all. Meanwhile, measurements of the corona's spectrum continued, during various eclipses spanning three decades. These established that the pattern of lines is due to an assortment of elements, and their nature further revealed that the source is exceedingly hot, with temperatures of millions of degrees. Thus early in the twentieth century it was finally confirmed that the corona is a dilute gas with an extremely high temperature, much higher than that of the sun's bright orb.

Having completed this history, my companion briefly reminded me that the early twentieth century was when perhaps the most famous scientific experiment in a solar eclipse took place. This was Sir Arthur Eddington's measurement in Principe on 29 May 1919, which proved Einstein's general theory of relativity.

According to Einstein, light is deflected in a gravitational field. The sun is the main source of gravity in the solar system and, if Einstein was correct, should deflect the light arriving from distant stars. During the total eclipse, Eddington found that the position of a star adjacent to the sun had moved slightly from its expected position. 'Slightly' is an understatement: it corresponded to a deviation of about ten parts in a million, that's like one centimetre in a kilometre. The amount was, according to Eddington, in agreement with Einstein's theory.

Subsequently people have debated whether Eddington's experiment was really as sensitive as believed at the time. And with all the rapture that accompanies totality, it seems hard to imagine anyone having the sangfroid to perform a delicate scientific measurement during such a singular event. I made a mental note that this was something to assess, once I had seen totality for myself.

* * * * *

He had just begun to tell me that solar and lunar eclipses are the tip of a very big iceberg—modern astronomy detects extra-solar planets when they eclipse their parent star—when my briefing on how science had exploited eclipses of both sun and moon was rudely interrupted: 'We are now boarding the flight for Lusaka', came the announcement. Passengers grabbed their belongings and began to head for the gate. As we made our own way down the corridors I examined my fellow travellers once more.

One man in the lounge had been especially noticeable during all this time. He wore a bushwhacker hat with eclipse medals around its brim, a sports coat that had been high fashion in the 1960s, and grey trousers that ended just above his ankles. With a nasal penetrating voice and an intensity that hinted at slight instability, he informed anyone within earshot the details of the eclipse, and of ones that he had seen before.

I played the odds, to convince myself that there were many places in Zambia to which eclipse chasers might head, so that it was unlikely that this was my first encounter with a member of our particular party, which would be less than two dozen strong. Now that the flight had been called and we shuffled up the jetway with our assorted equipment, the bushwhacker's voice trumpeted in my ear: 'I see you've got your tripod!' Unsure whether this qualified me as a professional, or a nerd, and now worried that any significant conversation on my part could identify me as his new best friend for the entire trip, I modestly made some non-committal remark and tried to remain anonymous.

The score of members of our package tour would all be seated near one another on the flight, so I breathed a sigh of relief when he moved to a different part of the plane to be heard, but not seen, again.

* * * * *

From London to Lusaka is a journey of 5000 miles, as the crow flies. By plane, it took some 13 hours, via Nairobi. The time change between Zambia and England was minimal, but we travelled from northern to southern hemispheres, from midsummer to midwinter. One of the hazards of long-haul flights, where you are crammed like sardines with minimal leg space, is deep vein thrombosis. A medical colleague had given me advice on how to reduce the risk: 'Before you get on the plane, drink a litre of water.' The reason, he explained: 'Then you will be walking up and down the aisle, to and from the toilets.'

Thankfully this advice has not been universally adopted. Nor do I bloat myself with water before boarding, but I do prefer an aisle seat, which allows me the opportunity for exercise. I walked along the aisle, eyeing the passengers, many of who were asleep. Books on astronomy and eclipses were prominent. I managed to avoid contact with the bushwhacker, but met several others who were on the communal quest.

I came across the professional once again, and suddenly remembered the question that I had forgotten to ask: 'What's the opinion about the biblical description in Joshua that in the solar eclipse, the moon or sun moved backwards?'

'Ah, you will only understand that when you've seen totality for yourself', he replied enigmatically.

6

'Who's Arranged this Eclipse?'

In the arrivals hall at Lusaka's Kenneth Kaunda International Airport, some twenty passengers from our flight were gathered around a huge man who looked like a relative of Idi Amin. He would be one of our drivers. Having flown some 5000 miles from London, we now faced a road trip of 200 miles, west and north of Lusaka, to our first destination—a camp in the Kafue National Park.

I greeted my fellow adventurers. Like me, they were all eclipse novices. I was relieved to discover that there was no sign of the bushwhacker or other evangelical eclipse chasers.

We made our way out of the crowded hall, to be confronted by an obstacle course of entrepreneurial porters. This motley crowd of locals wrestled our baggage trolleys from us, like muggers in broad daylight. They vectored towards buses, which carried the logos of luxury American hotel chains. 'No!' one of our group shouted, 'we're on an eclipse safari.'

With this discovery that we were not high rollers from the USA, the porters dropped us like hot bricks. They quit in mid forecourt, and disappeared in search of more lucrative punters. Meanwhile, we were rescued by Idi Amin's cousin. He led us into the heat of the car park and directed us to our transport. The three minibuses were distinctly lower quality than the luxury courtesy limousines to which the porters had originally headed, but were undoubtedly more suited to a safari.

With seven of my new companions, I squatted in one of these vehicles, my minimally padded seat located over one of the rear wheels, so that I had my feet on the wheel arch and my knees in my chest. Our luggage filled the well in the middle and the bench seat alongside Samuel—for that, I learned, was the name of the Amin doppelganger. Samuel occupied both the driver's seat and a considerable portion of the adjacent space. He turned to face his back seat passengers. With a cheery smile and a hearty 'Welcome to Zambia', he crashed into first gear and we lurched forward on the next stage of our adventure.

We had not even left the confines of the airport before it was apparent that this would be a rough ride. Suspension non-existent, the vehicle rolled like a boat as the shock absorbers had long gone. Once we were on the open road, it felt as if the van had square wheels. There would be 8 more hours of this.

* * * * *

We headed west along a strip of sweltering tarmac cut through scrub, bushes, and trees, which filled the vista all the way to the blue haze of the distant horizon. No lines denoted the middle of the road, no kerbs its edges. This was an African twentieth century version of a Roman road: two points, over a hundred miles apart, had been joined by a straight line, liquid tarmac poured over the stones, and some official had then decreed this to be a highway.

The route crossed interminable shallow valleys. We would rush down the gentle slope of one for a few miles, continue onwards across the base, and then slowly climb the far side. Smoke poured from the van's exhaust and its engine threatened to stall when any upward gradient grew steep enough to be classed a hill. Upon reaching the far crest, we would scan the new panorama, like mariners desperately seeking sight of land or, in our case, of habitation. Instead of novelty, however, like some video on a never-ending loop, each horizon seemed to repeat what had gone before: a ribbon of tar, whose parallel edges merged at a point in the remote distance. Another half an hour bumping across this never-ending terrain revealed this once far-away point to have a width of about two cars. Meanwhile, the vanishing point of perspective had now moved on a further 15 miles, to the hilltops at the far side of the next valley. And so it continued, as the road carried onwards into the heart of the continent.

Crammed and cramped within the minibus, we fell into a hypnotic doze. In my reverie I mused that this fulfilled my eclipse fantasy of childhood: eclipses are exotic both in reality and in their location. Surely no one would have chosen to make this voyage of purgatory were there not a special reason. Certainly I would never have done this but for the eclipse. A small voice in my head, however, kept saying: 'It'd better be worth it.'

I was brought out of my daydreams by the blast of a horn as our convoy pulled off the road and stopped. A track crossed the highway, and at this lone intersection was a ramshackle homestead, with rusting signs, which advertised Coca-Cola, Alka-Seltzer, and Capstan cigarettes. A primitive gazebo supported at its corners by stout branches of mopane trees provided some shade, under which stood a white-haired toothless man. As we clambered out of our vehicles, and stretched our limbs with relief, he encouraged us to purchase water, melons, and other fruit. I noticed the flies that already seemed to be enjoying these delights, and then joined the queue for the latrines at the rear, where lizards, frightened by our footfalls, scurried into the scrub.

We appeared to be in the middle of nowhere, yet within minutes several locals had materialised from the bush, laden with trinkets. They were eager to know who we were, where we were going, and why. On the side of the shack was a notice, which had been issued by the Zambian government and referred to the eclipse. It had a diagram of the eclipse track, and instructions to the population on how to behave, with advice to stay indoors in order to be safe. We pointed to this and explained our reason for being there. The old man, upon learning that we had travelled all the way from England to see this eclipse, went into a manic cackle: 'Just three minutes', he said, referring to the brief duration of totality. 'Just three minutes, ha ha ha.'

A small boy, no more than 12 years old, wearing shorts and a red Manchester United replica shirt, accosted me: 'Who's arranged this eclipse?' he asked.

Momentarily confused, I hesitated and echoed: 'arranged?' He explained: 'Is it the government doing it to make money?'

He was so young, but already wise in the ways of the world. 'No', I explained, 'it's a natural phenomenon. Sometimes when the moon crosses the sky it gets in the way of the sun.' He thought about this, his face showing doubt. 'If it's natural, how do you know it's going to happen?'

Now that is difficult to explain, not least when at a pit stop in the midst of the Zambian bush. I did my best to recall my own first encounter with an eclipse, half a century before, when I had been nearly the age of this young lad. So I repeated the gist of what my old teacher, Mr Laxton, had said back then. First, I said, the moon orbits the earth once every month. Once every circuit, the moon comes between the sun and the earth, but usually not quite in a straight line. Sometimes it's a bit higher than the sun, or a bit lower, but once in a while it's right in line and when that happens it gets in the way of sunlight.

'If you stand so that a mopane tree is between you and the sun, you will be in its shadow.' That was easy, because I was already doing that very thing, and the boy moved out of the sunlight himself, as if to act out the drama. So now we were set for the leap of imagination.

'If the moon is directly between you and the sun, you will be in the moon's shadow. The moon is very big and its shadow will cover from one horizon to the other. But the moon moves quickly and will be out of the way in a few minutes and it will be sunny again. For those three minutes, however, the sky will turn dark.'

He must have taken it all in, because instead of asking more about the 'why', he moved on to the when and how: 'How do we know this is going to happen here, next week?'

That was a good question, I thought, and far too difficult to answer well, but I had to say something to explain why we had come all this way. 'By keeping careful track of how the moon and the sun move across the sky, and doing arithmetic to check when they line up this way, it's possible if you're very careful, and don't make mistakes, to work out when this will happen, and where you have to be. You might be in the line of sight in Lusaka, but if instead you were just a few miles north or south you would see a sliver of the sun beneath the moon or above it. Only if you are at exactly the right place will the sun be completely blacked out. And the calculations show that this should happen, near here, about a week from now, for just three minutes.'

The boy listened. He was interested, but sceptical: 'I still don't believe it will happen, but if it does, then I will believe in science.'

As I re-joined the van, I mused that I was now older at the southern midwinter of 2001 than Mr Laxton had been at the northern midsummer of 1954. And, when the eclipse happened, in about ten days' time, I wondered whether the young Zambian lad too would 'believe in science'.

* * * * *

We left the highway and headed into the bush. The privations of our journey so far now paled into insignificance. The road from Lusaka had been a superhighway in comparison to the track upon which we now found ourselves. We bounced from one rut to another, across sand and scrub. Momentarily we were suspended in mid-air as the van dropped over the edge of what in the rainy season would have been the banks of a river. With a jolt we hit the solid mud, which had been baked by the sun to the hardness of primitive bricks. We were tossed from side to side as we lurched across the river's arid base, which was littered with the branches of trees and empty oil drums. Then it was on into the bush once more, as trees enveloped us. The trail, insofar as one could be recognised at all, was little more than a continuum of gaps between the trees, the result of elephants having bullied their path through the forest.

Suddenly the jungle thinned and we came to a vast plain, with intermittent bushes and baobab trees. The thick long trunk of the baobab rises for over 10 m without a single branch, until at the top there is a maze of tentacles, which spread like an open umbrella. The impression is of a tree that has fallen to earth from on high, but head first. Its top having plunged into the ground, its roots are now exposed to the sky, and radiate horizontally in all directions.

Our driver suddenly stopped the minibus, not so we could marvel at this most beautiful sight, but at a group of about a hundred gazelles. The animals were grazing while being watched by a leopard, which was stretched out in the sunshine, hidden in the grass.

One of the gazelles, stood slightly apart from the herd, appeared to have a translucent white plastic bag hanging from its rear. Through binoculars we could see that the spongy shape was in fact mucus, which surrounded a calf in the moment of its birth. The mother was on her haunches as her baby flopped onto the ground, where it lay, a lump whose legs thrashed around at random. The leopard rose ever so slightly, and stared at the new-born, gauging the chance of a meal.

The mother kicked the calf, at first gently and then more urgently, encouraging it to stand, to achieve within a minute what humans take many months to accomplish. Thankfully, we no longer have animal predators ready to devour the weak at the first opportunity, but in the bush the rule is survival of the fittest.

The pack was now aware of the leopard, and as the calf momentarily stood, only to slump to the ground once more, the group prepared to move off, away from the cat. The mother gazelle was in conflict, torn between protecting her offspring and ensuring her own survival. The leopard now saw its chance and began to stir, crawling a few metres nearer to the herd, before crouching once more and watching their every move. For us spectators there was the temptation to manoeuvre our vehicles, so that they would come between the gazelles and the leopard, to protect the calf. But that would have been to interfere with the natural order. The leopard too had its right to life and its survival required it to hunt and kill.

The drama of life and death on the grasslands of Africa would, within moments, reach its climax. Either the baby gazelle, having spent seven months inside its mother's womb and emerged into daylight, would soon be dead meat inside the stomach of a leopard, or like Lazarus would stand, to walk and run with the herd. And within seconds the calf achieved just that. As a child may take faltering steps after about a year, and gradually learn to walk and later to run, the gazelle—a mere 5 minutes old—found how to stand on its own four legs, gambol and leap, and enter the collective safety of the herd.

The leopard settled back into the grass, on the lookout for other prey. We restarted our vehicles and headed across the plain towards more forest, and camp at the junction of the Lufupa and Kafue Rivers. We spent a week there in the bush of the National

Park and then packed our bags, once more, in order to move to the far side of Zambia where we would intercept the path of totality: at Chiawa camp, located in a grove of mahogany trees by the Zambezi River.

* * * * *

Our small group filled the camp at Chiawa, which consisted of about a dozen thatched huts. We were on the north side of the Zambezi River, whose southern bank was the border of Zimbabwe. Mid-stream, a small island in no-man's land would be our base for the eclipse. From there we had a clear view along the valleys that the river had carved; the sun would be high above us, clear of overhanging foliage, and, perhaps most important, the surrounding water would stop us falling prey to lions and other wildlife, for whom the onset of darkness would signal night.

The eclipse would occur in the afternoon. So in the week that led up to it, we took time out after lunch to check the site, decide the best orientation for cameras and telescopes, and plan our strategy. There was great excitement about a week beforehand when the crescent of the moon could be seen rising above the distant peaks. With the sun already high in the sky, it was hard to believe that a week would be sufficient for the moon, barely above the horizon, to keep its appointment the following Thursday. The next day, however, the moon was well clear of the mountains at the appointed hour, and by Monday and Tuesday could be seen to have advanced across the arc of the sky, on course for its rendezvous.

By the Wednesday, just 24 hours before the moon's shadow would sweep briefly across the land, any intelligent observer could extrapolate the track of the moon and anticipate that the next day would be portentous. Did this play a role in how ancient witchdoctors predicted eclipses, I wondered? Even if it were not a total eclipse, the chance that the moon would cut a swathe across at least part of the sun was high, and with suitably Delphic utterances, the astrologers could cover themselves if it was but a glancing cut, while asserting proof of their magical powers if a more wondrous event occurred.

We had the benefit of modern science, however, and were assured that on the Thursday, when viewed from our location, the moon would begin its transit across the sun at about two o'clock. The sun would be eaten away for about an hour and a half, to be followed by 3 minutes of total eclipse, during which who knows what would happen. Then the moon would move away from alignment, as the crescent sun and daylight reappeared. This Wednesday lunchtime was our last chance to rehearse, to confirm the angles of our telescopes, the positioning of tripods, and to ensure that we would not trip over unseen hazards in the gloom.

The camp owners, an enlightened family from Zimbabwe, had thought of their staff, and made plans for them to enjoy the experience in safety. A score of plastic chairs were taken to the island and placed in a row, facing the sun. A set of eclipse goggles, filled with dark foil, had also been provided, one for each worker, so they could watch the crescent sun without harm (Figure 6A). They too took part in the dress rehearsal. Their boss told me that he would say a few words to them, to inform them what was going to happen the following day, and then as I was a scientist he would invite me to explain what causes a total eclipse, and to give them

Figure 6A. Eclipse goggles in Zambia (photo: Frank Close). See Plate 8 for a colour version.

a foretaste of what to expect. He would translate my words for the benefit of all.

The boss started an oration in the local dialect, which went on for some time. This was because some of his audience kept interrupting. During these interventions they waved their arms and appeared to argue. After an interminable time everyone seemed satisfied, and it was my turn. But first I asked the boss what had happened so far.

He told me that some of his workers initially refused to take part. They wanted to stay indoors, afraid that evil spirits would be released during the eclipse. Three of the men asked to return to their families in the village, several miles away, so that they would be present if the spirits of dead ancestors returned to their homes when the sun was taken from the sky. The boss had told them that there was nothing to worry about; that the stories about evil spirits, disasters and mayhem were rumours put about by bad people. Their families would be safe, he assured them. He added that they would be safe too, more so on the island than in the bush-camp. And he could promise this to be true because in the camp they had 'a man who has seen many total eclipses, a man of knowledge who will now explain everything about what will happen and why you have nothing to fear'.

No pressure there, then! My 'many' eclipses amounted to just one, and that had been beneath cloud. There would be no danger of that on this occasion as we had not seen a single cloud all week.

I was now due to explain the eclipse to an expectant audience. My mind went back to that demonstration in my childhood, now 47 years in the past. I had no football, cricket ball, or lantern, however, and the bright natural sunlight of Africa was far from the sombre theatre of St Mark's School, Peterborough. So, instead, I adopted the style of the Royal Institution Christmas Lectures, in which one uses the audience for the demonstrations. The camp's owner played the role of the sun, I was the moon, and two of the staff played the earthlings. One of them pretended to be an observer in Lusaka, while another represented the camp where we were. I walked slowly between them and the boss, illustrating how the moving moon would in turn block the view of the sun from each of them. They all agreed that they understood, so we moved to the next lesson: how to look at the sun through their goggles.

I put mine on, and held one earpiece gently in my fingers as insurance lest a puff of breeze disturb the spectacles while I looked at the sun. Then I looked up until I found a circular bright globe, a white dot in the middle of my vision. They each did this for themselves. The boss emphasised how important it is to use the goggles, and to hold them securely. I explained that this would be essential if they chose to look at the sun during the first 90 minutes, while the crescent sun was still bright, but that during totality itself it would be dark and safe to look with their naked eyes. We promised that about a minute before totality, when a mere sliver of the sun remained visible, someone would shout to say when it was safe to look. And as the end of totality approached, a further warning would be given so that they would replace their goggles as the flash of the new-born sun appeared. There is no point in wearing goggles during totality itself, of course, as the sun is obscured, and the sky is dark. All this was explained on the Wednesday, and repeated again on the day itself.

At the actual eclipse, everyone carefully put on the eclipse goggles. Some of the staff, however, never removed them. What sights they missed.

* * * * *

Thursday, 21 June, midwinter's day in the southern hemisphere, we were up before sunrise. The schedule for the morning was to have a walking safari through the bush around the camp, accompanied on foot by a guide with a gun. We walked about four miles, admiring termite mounds, animal trails, and carcasses of animals from the previous night's kill. A fascinating adventure, but our minds were elsewhere. The eclipse would begin in a couple of hours and we wanted to be back, and ready for it. As we returned, however, we discovered that a herd of buffalo had straddled the trail and blocked our route home. We stopped, thankfully downwind of the animals, where they would not pick up our scent, and waited. Slowly, one by one, more than a hundred buffalo meandered across the track and rambled tortuously away from us.

The thought that we might have been trapped together with the herd in the gloom of the eclipse gave added urgency to the wish to get back to camp. This we did without further alarms. Then, having showered, and gathered our equipment, we prepared to get in the canoes that would take us to our island observatory.

Having been isolated in the middle of nowhere all week, we were now aware of company. We had heard the sounds of light aircraft late the previous evening, and again in the morning early, at sunrise. Trippers, on what was literally a flying visit for the day, now became apparent. One of their buses had become stuck in a tributary of the Zambezi, within view of our camp. A herd of elephants had become interested. The party, dressed more for tea at the Ritz than a day in the bush, and used more to seeing elephants at the zoo or in a parade at a circus perhaps, seemed unaware of the potential danger. One man in a suit and tie, accompanied by a well-dressed lady, who was tugging her high heels back on, could have passed for characters in a novel by P.G. Wodehouse. The owner of the camp, together with some assistants armed with guns, went to their aid, managed to extricate them from their self-made mess, and helped them on their way. Our relief was as much for their safety as the selfish one that the horde would not overrun our careful preparations.

But as we were about to paddle the few metres to midstream, a new hazard appeared. Small eddies in the water revealed the hidden presence of hippos, lurking beneath the surface.

Ever since Flanders and Swann immortalised hippopotami wallowing in the mud of the river, their bodies cooled in the water with only their nostrils visible above the surface, these creatures have had a romantic image. Yet in reality they can be extremely dangerous. While crocodiles are infamous for grabbing unsuspecting tourists who venture too close to the river, or even foolishly dip their toes into the water, it is even more hazardous to paddle a canoe over a submerged hippo. The beast rises out of the riverbed, the canoe and occupants are hoisted into the air, and then dunked in the water. If you are lucky you reach the shore before the crocodiles eat you, or the hippo tramples you. In total, more people are killed by inadvertently disturbing a hippopotamus than by other encounters in the bush.

With our canoes in convoy, we followed the leader as he set off upstream, beyond the hippos and then circled round and let the natural flow bring us safely to shore. At last, we were safe and ready for the show.

The contrast with Cornwall could not have been greater. The sun sparkled in a cloudless azure sky. From our perspective in the southern hemisphere, it moved gradually from right to left above us. As gravity's inexorable force spun the moon into position, out in space its shadow shot out like some negative beam from a lighthouse, eventually to be lost in the black void of space. Except that at dawn on 21 June 2001, at a point about 2000 miles away from our bush camp, the earth's orbit had intercepted this pencil beam of darkness. Now, as we gathered that lunchtime, the earth's surface was spinning us towards the moon's shadow. From our perspective, unaware of the earth's rotation, the shadow was rushing towards us across its surface from west to east at a mile every second.

About an hour before the main event was due, someone shouted 'first contact', which is astronomer-speak for 'the moon has begun to cross the sun'. Through binoculars, suitably protected against the glare, a small nick could be seen disturbing that perfect circle, and it was growing. For me, this is one of those moments where I feel humbled by the ability of science to predict: on this day, at this particular time and place, the moon will begin to be in direct line of sight to the sun. Newton's laws of motion are inviolate; nothing could now change what is inevitable. No quirk of meteorology could obscure the view. Like lovers, when they know that they are about to lose their virginity, we too were full of excited expectation, certain that something wonderful was about to happen.

* * * * *

As a disc of pure blackness began to slide across the face of the sun, dusk began to fall. But it was a strange twilight. Turtle doves began flying low across the trees, and vultures coming in to roost circled lower and lower, like at a normal sunset, except that at the onset of totality, darkness was so sudden that the vultures landed in the dark. Hippos began to leave the river for dry land, right where our cameras were set up. A few rifle shots into the air from our guards encouraged them to leave. It seemed like a normal evening for the animals and birds, but for us humans it was strange: the light got dimmer but the shadows didn't lengthen.

Our shadows took on a split personality, showing strange bifurcations as the crescent remnant of the fast disappearing sun illuminated us wanly in the deepening gloom. As totality approached, there was an intense sense of anticipation. The air cooled, and then in the west, a wall of darkness, like a gathering storm, dimmed the bright blue sky: the moon's shadow rushed towards us. No wonder the ancients were terrified. In an instant we were enveloped by the darkness as the last sliver of the sun disappeared and a black disc appeared in its place, quivering in front of a supernatural pallid glow.

I had waited so long for this moment, that when it happened, it felt unreal, like a dream, as if I was a spectator at someone else's eclipse party. Slowly my mind adapted to the unworldly sight: the black hole is the real moon, obscuring the real sun whose coronal glow, now visible, was like some desperate attempt to remind us that the source of life was still out there, and would not leave us. Then I broke free from this reverie. Aware that I was shaking with emotion, I just let the experience take over.

The moon's shadow brought night to the dome of sky overhead. If you looked directly upwards you could see stars like it was normal night. But in the midst of the night was an awesome sight, the stuff of nightmares. A circle of profound blackness, a veritable hole in the sky, was surrounded by shimmering white light. A botanist in our group saw it like 'a black sunflower with the most delicate of silver petals'. One of the Zambian staff, who had removed his goggles, asked: 'Is that the eye of God?' A colleague also felt similar emotion: 'It was like looking into the valley of death, with the lights of Heaven far away calling for me to enter.'

The contrast was like marble that is floodlit in the darkness of night. Then as one shifted one's view further, the night was revealed to be only in a dome above us, as if floating on a purple sea which in turn rested on a 360° sunset. It was as if we were witnessing the end of the world; the depths of infinite space were hovering above. It was simultaneously ghastly, beautiful, and supernatural.

Even for a humanist, the vision was such that I thought, 'If there is a heaven, this is what its entrance is like.' The heavenly vision demanded music by Mozart; instead we had the crickets. Crickets had started chirping up to 10 minutes before totality and continued throughout, aided and abetted by the deep 'ho-ho' of the hippos and a chorus of frogs.

By contrast, Momba, the tame hornbill at Chiawa, appeared oblivious to the spectacle and spent her time trying to peck the shiny protective spectacles from some of our heads.

The eclipse was predicted to last three and a half minutes, and my watch confirmed this, yet emotions were so charged that I could swear that time had stood still and the whole event had lasted less than a minute. Then, when the 'diamond ring' effect lit the edge of the moon, daylight returned in an instant. That is the critical moment when one must avoid direct sight of the sun. After 3 minutes of darkness, with pupils enlarged, the sudden brightness can be literally blinding.

A wall of darkness rushed away to the east to be replaced by the splendid light of a sunny day. In that moment wildlife took up its day routines, carrying on as if waking from a catatonic state, as if nothing had happened. Seven hornbills flew out of a tree where they had been roosting only moments earlier. Normal life returned.

* * * * *

As we approached England on the flight back to London, I found the professional astronomer standing in a galley. He asked what my reactions to the eclipse had been, and I told him much that I have written here. Then I added a thought that until that moment hadn't consciously occurred to me.

'I hadn't anticipated that because we were in the southern hemisphere that everything would be back to front. It was like watching the sun and moon in a mirror.'

Whereas in England we watch the sun and moon, in the south, move from left to right during the day, they were to the north of us in Zambia. This made the sun rise on the right and set on the left, which meant that during the day their direction of travel was from right to left across the sky. I had been prepared for that, but hadn't foreseen the full extent of the implications for the eclipse itself.

As the moon crosses the sun from west to east, this means that the sun converts initially from a circle to the shape of the letter C, when an eclipse takes place in the northern hemisphere. In Zambia, however, the moon traversed the sun from the left.

My companion laughed, and then made a confession.

'Yes! We nearly overlooked that, ourselves. It was only six months ago that someone suddenly pointed out that totality would develop from the left to right, and we had to redesign part of the experiment!'

I suddenly remembered his enigmatic remark about the retrograde movement of the sun and moon as recorded in Joshua. I had been so overwhelmed by the diamond ring and shimmering lunar silhouette that I had forgotten to watch for this, so I asked him to explain. But he had hardly begun when the flight attendants told us to return to our seats in preparation for landing. He parted with a yet more obscure remark: 'When you see an eclipse with some clouds, that's when the illusion works best.'

I returned to my seat, this adventure nearly over. Next time—for now I was hooked, and was sure that there would be a next time—I must look for Joshua's lunar illusion, especially if there were clouds.

Where would the next opportunity be? By chance, a year later on 4 December 2002, Angola and Zimbabwe would again host totality, but only for 1 minute before the path crossed the Indian Ocean, far from any land. The latter was impossible, and I had no urgent wish to return to the same region of Africa so soon, and for a short-lived eclipse. The year after that, on 23 November, would require a trip to the Antarctic. The greatest length would be under 2 minutes, and to see even that would require a trip into the heart of the continent. It would be possible to take a journey on a ship, to within sight of Antarctica, and if the seas were not too rough, and the skies clear, to experience a minute of totality.

I was not yet an addict, and decided that I could wait for longer. In 2006, totality would cross the Sahara Desert in Libya. That sounded exciting, and there would be little likelihood of clouds there. As for my hope to solve the enigma of Joshua, at least now I knew what to look for, and also what a total solar eclipse is like. Except, as others had told me, and I was about to discover for myself: every eclipse is unique.

7

Earthshine in the Sahara: Libya 2006

On 29 March 2006, the path of a total solar eclipse was due to cross Africa from Ghana and Nigeria in the west, through Niger, Chad, and Libya in the centre and north, before exiting to the Mediterranean at the border with Egypt. It would then cross Turkey, where Tivas, the highest city in Central Anatolia, would be on the path of both this and the 1999 eclipse. One possible venue, to which several prospective expeditions were headed, was the coast by Egypt. But for me, eclipses are the icing, which calls for a cake, and in this case the substance would be the Sahara and Libya. The best guarantee for clear skies would be in the heart of Africa, but to reach remote parts of Niger, Chad, or the Sahara Desert of southern Libya, where the eclipse would last the longest, presented a challenge. The path was hundreds of miles from the nearest local airstrip, let alone a recognisable airport with a connection to any major centre.

This was a reminder of how vast the earth is, and how solar eclipses can be hard to find. On the globe there are vast expanses of central Africa, which are undistinguished on satellite images, with no railway line or metalled highways. Imagine a journey from Stockholm in Sweden to Marseilles on the Mediterranean coast without encountering any settlement larger than a village in the whole trip. Areas the size of England or France could be superimposed on a map of north-central Africa without encompassing a single major town, or even a tarmac road.

Even were it to include such a road, this ribbon link between two cities thousands of miles apart would probably have no significant crossroads throughout its entire length. But across such a swathe, the eclipse track would pass from south-west to north-east and in so doing cross a handful of roads that connected Tripoli or Benghazi with southern Libya. This at least offered a way into the heart of the desert and access to the eclipse.

The border of Libya and Chad, where the eclipse would be greatest, was lawless. Although nominally on the trans-African highway from Tripoli to Cape Town, this route is a concept more than a reality. First, 'highway' is a relative term. This is no multi-lane trunk road, not even a dual carriageway, but instead is a strip of tarmac, which weaves across the sand dunes and is just about wide enough for two narrow vehicles to pass one another with care. The distance from Tripoli to the south of Libya is about 1000 miles. The road surface ends at least 300 miles before the border, after which it would be necessary to follow mere tracks across the desert. This was clearly out of the question for all practical purposes.

Another way to intersect the eclipse path looked more promising. About 500 miles south-east of Benghazi is Al-Jawf, a town of about 17,000 inhabitants. In the vast swathe of the Sahara, Al-Jawf is a major centre, linked to Benghazi by a solid road. Some 300 miles south of Benghazi, this highway would intersect the eclipse path at a point 28° 14′ 3″ north of the equator, and 21° 29′ 25″ into the eastern hemisphere.

This previously undistinguished grid reference became a target for eclipse chasers. I discovered that at this spot some intrepid adventurers planned to set up a temporary encampment of tents, which they had grandiosely named 'Eclipse City'. The average temperature in the vicinity of Eclipse City during late March is above 90°Fahrenheit and the annual rainfall negligible, less than a millimetre. Satellite images of the site taken on 29 March over a 15-year period showed clear skies every time. The total eclipse was predicted to last for 4 minutes and 4 seconds, and would be the first occasion in daytime that the sun had not blazed uninterrupted on any 29 March in recorded history. We had found an ideal spot to experience an eclipse. Now it was a matter of organising how to get there.

Then I had a stroke of luck. I discovered that groups of astronomers had charted a fleet of buses from Benghazi to the centre of the eclipse path, and to make the venture more appealing still, had combined it with a Mediterranean cruise, from Crete to Tripoli, with visits to the historical sites of northern Libya, then to Benghazi and the Sahara Desert. The particular group with which I would travel planned to make the rendezvous in comparative luxury. Instead of camping in the arid heat of the desert for several days around the eclipse, we would arrive in Benghazi by ship, and before dawn would drive into the desert in time for the midday rendezvous. We would then return to Benghazi late that night. This seemed to be a perfect fit with my strategy that eclipses could be an entrée to places that otherwise I would be unlikely to visit.

* * * * *

Libya isn't all desert. The Mediterranean coastal fringes are hilly and green, with fruit trees in abundance. This area is pleasantly temperate, in contrast to the dry heat, which these modest mountains trap within the continent. As one leaves the slopes of the hills, and moves inland, the vegetation disappears. Scrub and wasteland, with occasional bushes and lichens, soon give way to a desert, first of dirt and dust, and eventually the classic wastes of light beige sands of the Sahara. It would be somewhere in that arid interior, a land of nomads, camels, and sporadic oases, in which we would meet the moon's shadow.

Would we be alone? How many others would we find camped in the wilderness? Libya is a vast country, with a population only of six million, most of whom are concentrated along the coastal fringes. There seemed little danger of hordes of inhabitants descending on the venue, unlike the case of Europe in 1999. On the other hand, if the locals did decide to watch the event en masse, there was only one practical route for vehicles to reach the path overland, and that was by the highway from Benghazi to Al Jawf.

How many people with a common goal could the road accommodate? I made a rough calculation, and the answer didn't look promising. First, Libya was not geared up for tourists, and it appeared that the needs of a thousand people on our ship had already commandeered most of the available coaches. To move 100,000 Libyans deep into the Sahara would take more than

20,000 cars. Line those up at 2 m apiece and you have a train 40 km long, bumper to bumper. Once these were on the move, with a reasonable gap between successive vehicles, the result would be a rush hour traffic jam stretching from Benghazi most of the way to the site (Figure 7A). Even if only a fraction of that number were to set out, there still seemed plenty of opportunity for mayhem. Had the organisers factored this into their plans?

* * * * *

On the day before the eclipse we all gathered in the salon of the ship for a briefing. When the question of possible congestion on

Figure 7A. Saharan traffic jam: rush hour en route to the path of totality in the Sahara. Expect traffic jams in the USA 2017 to be even more extreme (photo: Frank Close). See Plate 9 for a colour version.

the road came up, the tour guide confidently assured us that we would have no problems reaching the site. With an anachronistic colonial perspective, he naively believed that most Libyans were unaware of the impending eclipse, and that few of those that were would have the inclination or the transport necessary to make the trek. Nonetheless, to be safe, we would receive an alarm call at 2 a.m. for a three o'clock departure on buses, filled with petrol, ready to get us to the eclipse rendezvous with plenty of time to spare.

All of these predictions soon turned to Saharan dust.

* * * * *

A journey to an eclipse is more than a vacation; it is an expedition. Why else would one accept the need to gather by three o'clock in the morning for a minimal breakfast of a croissant and coffee before disembarking to gather on the dockside in the middle of the Libyan night? Immigration officials, concerned lest some passengers chose to migrate to Gaddafi's fiefdom without the necessary permission, instead of returning to the comfort of their homes in California or the Cotswolds, were ready for us.

We were lined up as if on parade, to be counted like prisoners of war. Then we were gathered in small groups, and guided towards a convoy of buses.

More than 20 assorted vehicles awaited us, on acres of tarmac and weeds, flanked by warehouses built of crumbling concrete and rusty girders. The scene was as if time had stopped, decades ago. A dozen of the buses would have been classed as luxurious in the 1950s. Familiar to those of us who had grown up in the Britain of that time, their rounded fronts and streamlined frames were the pinnacle of post-war sophistication. Their logos informed that they were now based in Egypt. Sometime before the Suez crisis of 1956, it would seem that British Leyland had exported the vehicles to Egypt where, half a century later, they were still in action, on loan to Libyan adventurers.

Of a more modern vintage, but of basic build, were buses of a military shape. Had they been painted yellow, they could have doubled for North American school buses. Finally, there were three superb modern coaches, from Tripoli, which would have graced any tour company in Europe or North America. Rumour was that these were part of Gaddafi's personal collection, his contribution to ensure a good press. We were, apparently, one of the first large tours to come to Libya in recent years, and were seen as a new beginning of tourism.

Those of us near the front of the column had been directed towards some of the more basic vehicles, but we conveniently 'misunderstood' instructions and made straight for the more splendid trio. They even had toilets, which on a journey of 300 miles across the Sahara seemed more of a necessity than a luxury. We flopped into our seats, settled in to sleep, and waited. A guard entered the coach, stood at the front of the bus, counted us, and left. And so we continued: stationary.

At last, after an interminable time, the convoy began to move. We crossed the dockside at walking pace. At the end of the row of warehouses we turned left, then round a corner to the right, and stopped once more.

Our coach was third in the convoy. This gave us a good enough view up front to identify the reason for the hold-up: 20 m in front of us the gates to the docks, the barrier to our entire adventure, were bolted shut.

There was no sign of any guards to unlock them, or of any initiative to look for anyone with a key. A guide arrived at the bus and announced that we were awaiting security police. Only now, as we prepared to venture into Libya at the dead of night, did we learn that armed guards would accompany the convoy as protection against terrorist attack. Today this news would be no surprise. In 2006, although it was five years after the horror of the Twin Towers, such risks were not yet part of daily affairs, especially in Libya, closed to the outside world and under Gaddafi's strict regime.

If you have ever been trapped in a stalled elevator, there is a short period of a few minutes during which you do not panic, as yet unable to accept that this is really happening. So it was with our halt at the dock gates. Initial amusement at the poor planning, began to give way to worry that we could be trapped there while a solar eclipse passed through Libya, 300 miles south of Benghazi.

After perhaps 5 minutes, though it felt longer, at last someone appeared, and opened the gates. The convoy crawled out from the docks. Once we were outside on the street, a group of police on motorcycles, and military in two armoured jeeps joined us. We headed off into the sleeping city, en route to the eclipse without further ado. Or so we thought; in reality, this was just the start of our frustrations.

* * * * *

Two days previously we had visited Tripoli. We had seen apartment blocks, maybe five storeys high, typical of any international city. In capitalist countries these are used for publicity or information. In New York, London, or Beijing, for example, tall buildings are used to host gaudy neon-lit commercials for soft drinks, cars, and electronic goods. In Moscow, during the 1970s, they carried political slogans: propaganda for Marxist-Leninism. In Tripoli of 2006, what at first glance appeared to be a promotional advertisement for the latest Hollywood blockbuster, with its handsome star displayed 20 m tall, turned out to be a ten times life-size image of Muammar Gaddafi.

These Gaddafi giants were common around the city, and came in several forms. Some displayed him in a quasi-military uniform more suited to a character in a Gilbert and Sullivan opera than a serious army; another showed him in the flowing robes of a Bedouin chieftain; in yet others he wore a tracksuit, as if he were a sporting hero. But all shared one common feature—he invariably wore designer sunshades.

Whether Benghazi also sported Gaddafi modelling sunglasses was hard to discern. Dark and empty at three in the morning, there were few signs of life and minimal illumination in the streets. If Gaddafi adorned the high rises, it was either too gloomy to see or I was too tired to care.

We drove along a wide boulevard, across a major intersection into a lesser street. This was especially gloomy, but we stopped in an area where there appeared to be some illumination, at which point the driver turned off the engine. For a few minutes nothing happened, until the door of the coach slid back to reveal our guide once more. The saga of earlier was repeated. The news this time was that we were at a petrol station, where the convoy would take turns to fill its fuel tanks at the solitary working pump.

Libya was a country awash with petrol, with grand designs for pipelines that would bring oil from the desert to the coast, and transport water inland to 'make the desert bloom'. Petrol was almost as cheap as water, yet—we now learned—the buses had passed the night with their tanks almost empty, lest anyone should attempt to syphon their contents. So our convoy of vehicles, which we had been promised would be filled and ready to go, had instead taken a diversion to visit Benghazi's 24-hour gas station. One by one, each vehicle now took its turn to fill with enough petrol for a 600 miles round-trip journey into the desert.

At some point, a second pump must have started, as otherwise we would have been stuck there for longer than the precious half an hour that actually elapsed. Finally, 2 hours after setting off, and little more than 5 km from our starting point, we headed off once more, with our security outriders and flashing lights attracting maximum attention. We even began to travel at a reasonable speed as we weaved through the suburbs and out of Benghazi onto what promised to be an open road to the south.

By 6 a.m. it was dawn, and Benghazi was behind us. We watched a red glow in the east spread under a turquoise sky, with a thin crescent moon visible too. Moments later, a throbbing ball of light and heat rose from the sandy horizon. Even at dawn, the sun felt like the blast from a furnace. It was hard to believe that the moon—whose surface area is no bigger than Africa—could ever black it out.

For most Libyans it was probably just the dawn on an average day. We passed through several small villages, whose tiny houses seemed to have been dropped onto the baked mud at random. Shutters on stores were being opened. Men gathered to smoke, and talk at every small square in the townships; men drank coffee to start the day; men pushed bicycles or walked among the rubble at the roadside, while avoiding dogs which scavenged for scraps. Street life in the early morning of a Libyan village seemed to consist of adult males and dogs. Women and children were all but absent.

* * * * *

Then the villages were gone, and we were in barren desert. The view was of sand in all directions, but for a scar of tarmac and occasional fences of barbed wire. These strips of wire, whose presence had no plausible explanation other than that they were 60-year-old relics of the world war, were littered with polythene bags, which had been thrown from vehicles and blown in the wind.

With no trees or vegetation to provide a barrier, the wind blew the sand without interruption. We could see its grains in motion, as they merged into small mounds, which obscured the highway. Its tarmac surface would be frequently lost to view as it disappeared into the desert sand for a hundred metres or more. Drivers had to trust to luck or judgment as they interpolated the path that linked the point where the dark strip vanished to that where it emerged, way ahead. Having then set their vehicle on this course, they had to hope that the trajectory included a solid base beneath the sand.

The road was built for trucks, whose drivers assumed that it was theirs by right. Car drivers have to be prepared to take avoiding action, which means that they must leave the road, and take their chance on the hard mud, or soft sand, alongside.

I glimpsed a distant oil refinery, and one township, which had grown around an oasis. Otherwise the view from the bus for the next several hours was of nothing but sand dunes. Sand, and more sand, until suddenly I saw what appeared to be the waves of the Mediterranean breaking on a sandy shore about 200 m from the road. I had never imagined mirages to be so dramatic. I could well imagine that stories of lost travellers, going mad as they rushed towards apparent oases in the desert, might be true. The tempting ocean waves kept their distance from us, forever out of reach like the end of the rainbow.

We met occasional trucks, which were heading north away from the eclipse track. A couple of cars overtook us, travelling south. These apart, we saw no other sign of life until about 9 o'clock when we became aware of occasional cars and motorcycles racing alongside and then overtaking us. Their numbers gradually increased, and we realised that, contrary to our guide's optimistic assurance, large numbers of Libyans had the same idea as us.

What had happened was that after their morning prayers, eclipse-savvy citizens of Benghazi and its surroundings had climbed into cars, or mounted motorcycles, and flooded the desert in a tsunami of high-speed vehicles. After some 3 hours this wave of traffic had caught up with our convoy, which was advancing at a more leisurely speed of about 50 miles an hour. The hordes now had the challenge of passing our convoy of buses, which were spread for over a mile along the highway.

Motorbikes slalomed in and out, overtaking one bus, undertaking another. Cars quit the tarmac and trespassed onto the sun-baked earth, which paralleled the road and was hard as rock. But for the more impatient, adventurous, or reckless among them, the highway and its adjacent stony ribbon weren't challenge enough. Like a jet of water released from the nozzle of a fire-hose, vehicles sprayed onto the sands beyond the solid periphery. Clouds of dust, like a sudden sandstorm, obscured our vision. The buses slowed as the wave of cars, jeeps, bikes, swept past us.

Skilful drivers of off-road vehicles managed to re-join the highway ahead of us, and raced onwards towards the south. But for family saloons, the Sahara Desert can be a trip too far. Like the fable of the hare and the tortoise, our ambulatory buses soon overtook cars that were stranded at the roadside, or littered the desert, stuck up to their axles in the sand. The drivers of more robust vehicles stopped to help their beached colleagues, and made desperate attempts to dislodge them. Young men dug away excess sand from the wheels, put carpets underneath the tyres, and pushed hard. Some attached towropes between stranded saloon cars and better-equipped vehicles, which had a better grip on the hard shoulder of the road. Our convoy, meanwhile, rolled on at a stately 50 miles per hour.

The tarmac cut a path through sand dunes, which were humped in ridges, hundreds of metres in length and up to 10 m high. They looked like the swell in the middle of an ocean. Winds had eroded these mountains into an anonymous patchwork of shapes. Like fractals, there were small hillocks and dips, larger mounds and hollows, all within the larger scale ridges and troughs, mountains and valleys. A mathematical quilt of self-similarity, its patterns repeated on all scales of size, which gave entrancing sculptures and yet random anonymity. Every dune was unique, yet undistinguished. Hills and hollows filled the foreground, stretched to the far horizon, and then presumably carried onwards for a thousand miles or more. How would we know which crest marked the path of the eclipse?

For more than 3 hours we passed through this seascape of sand, with no markers along the road to show how far we had come, nor how far remained to travel. We had no real clue as to where we were. A soporific sense of being lost in a groundhog day of never ending repetition made what happened next seem bizarre. A few metres from the road, atop a dune, someone had erected a wooden fingerpost in the desert sand. On it, in brown paint, they had written in English: 'Eclipse site 1 km.' Beneath this, the message was repeated in Arabic.

At first glance, the sign merely pointed towards another ridge of sand, higher than the others, beyond which we could see nothing. Then, as we crested the rise, we saw a second sign, which announced that we were within 500 m of our goal (Figure 7B). As if in confirmation, we could see scores of people on the far side of the plane, their blankets and bags on the sand, with impromptu sheets strung to give some shade from the heat. Dozens of tripods were

Figure 7B. 'Solar eclipse watching point 500 m' (photo: Frank Close). See Plate 10 for a colour version.

spread across the desert like miniature oil derricks. They carried all manner of telescopes, cameras, and other optical instruments. Plastic bags, filled with sand, provided weights to stabilise them.

I concluded that either large numbers of people had made the same miscalculation, or we had arrived at 28° 14′ 03″ north, 21° 29′ 25″ east, the point where the moon's shadow would cross the Al-Jawh highway. We had about 90 minutes in which to prepare for totality (Figure 7C). First contact, where the moon begins its transit across the solar disc, was only minutes away.

* * * * *

Our convoy of buses pulled off the highway, one by one, and aligned themselves on the hard clay shoulder. Freed at last from our prisons on wheels, we walked about 100 m into the desert, so

Figure 7C. Preparing for totality (photo: Frank Close). See Plate 11 for a colour version.

as to get some feel of its isolation, and to have an unimpeded view to both the south-west and north-east along the eclipse path.

We also managed to rendezvous with various friends who had joined us for the expedition.

Peter Boldon, who had watched the 1999 eclipse from the restaurant on the Cherbourg peninsula, was with us, along with his wife and two of his daughters. He already knew that this experience would be a legacy to pass on. My sister-in-law and her husband, who had been brainwashed by my anecdotes for years, had decided to come and see for themselves. Finally, Bill and Cathy Colglazier, two friends from our time in California and the total lunar eclipse of 1971, had come from the USA seduced by my enthusiastic descriptions.

They had travelled all this way to Africa, via Europe, to witness the eclipse totally trusting my assurance that this would be the experience of a lifetime. To achieve this rendezvous, still somewhat jetlagged, they had been dragged out of their bed at two in the morning, and then spent 9 hours in one of the most basic buses, which had no air conditioning and no toilet. Bill and Cathy located us among the throng, and trudged across the sands in our direction. They looked hot, thirsty, exhausted. At that moment, our friendship of three decades was facing its greatest test.

Bill saw me, dropped his bags on the hot sand, and looked at the sky.

'Frank: This had better be good!'

* * * * *

The temperature was already 40° centigrade, over 100 Fahrenheit, but it was tolerable because the air was dry. We laid out sheets to present some protection from the sand, which was scorching hot underfoot, and attempted to create some form of shade with umbrellas that our guides had thoughtfully provided.

There was not a breath of wind, except within the immediate vicinity of the ground. Hot air rises, and the sand passed accumulated solar energy upwards to the air by convection. This subtle motion of the atmosphere gave a shimmer immediately above the surface, which, together with refraction of rays from the blue sky, produced an eerie vision of lagoons in the hollows. The illusion of water on the sand, a mere 10 m away, was utterly realistic. This mirage, together with the people, towels, and sunscreens, could have doubled as a scene at the Riviera during the height of the summer vacation.

We plastered ourselves with factor 50 sunblock, and covered our heads with wide-brimmed hats. I held up a piece of welder's glass to shield my eyes as I took my first look at the sun. The thick ceramic sheet looked opaque, but in the middle of the black background the sun formed a bright circle, tinged with a greenish hue by the glass. At first sight this appeared to be a perfect circle, until closer examination revealed a trifling imperfection towards the lower right of the rim. Within a moment, this small dent had grown enough to be revealed as a tiny arc of darkness eating into the solar disc. This confirmed that the moon had made 'first contact'. The eclipse had begun.

This announcement brought with it, as always, that moment of wonder: science can predict the heavens, such that on this day, at this moment, from this unique perspective, the moon will begin its transit across the face of the sun. For the sixteenth year in succession, there was not a cloud in the sky over this hitherto unremarkable spot in the desert. The bus drivers opened up the luggage compartments of their vehicles, in order to find some shade. Then they climbed inside and lay down for a nap. They had driven for 300 miles, and would have to make the return later that day. For them, sleep was more inviting than the chance of watching a solar eclipse.

* * * * *

As the moon edged slowly across the face of the sun, the temperature gradually fell. Initially the change was barely noticeable, but as high noon approached, and with half of the sun by now obscured, the desert heat had softened to that of a pleasant summer's day, probably about 25°C. Totality was about half an hour away.

A view from the International Space Station would have shown a million square miles of pastel shaded desert sand, bathed in the direct glare of the midday sun, with Eclipse City inside the penumbral disc, which covered about 10,000 square miles (See plate 2). Although this was trifling compared to the whole of the Sahara, its ambient temperature was some 10–15°C lower than the surroundings. This already began to affect the climate. At the centre of this shadow, still more than 100 miles away from Eclipse City, was a dark circle—the 'umbra'—a patch of night about 30 miles in diameter, where the temperature was relatively cold, more akin to that of a March day in England than the Sahara. The cool air in the umbra sinks, while warmer air in the penumbra rises, but not as rapidly as the hotter air in the fully illuminated desert. A microclimate was heading our way, bringing with it a possibility that we had not previously considered: would a desert sandstorm obscure the sun, like some re-enactment of the extended darkness at the crucifixion in 33 AD?

This thought was born after the initial excitement of physically sensing the approach of the impending climax of the eclipse. The temperature had fallen, adiabatically, and the atmosphere at large had remained still, so far. But with about 15 minutes to go, and the sun now like a thin capital letter C, we felt puffs of cool air on our skin. Next, we saw occasional dust devils, spontaneous vortices of sand, which would erupt as cool air cascaded downwards to replace hot air that was rising from the scalding surface. Having sunk, this cool air would in turn be heated by the desert sand, which was still too hot to touch. The newly heated air would then rise upwards. As water cannot drain down a plughole without rotation, nor does air rise and fall in smooth laminar flow. Instead, small wisps, inverted cones of sand no more than 2 m high, and less than a metre at their greatest radial extent, were whipped into motion. They would spiral for a few seconds and then die out.

This eerie presentiment of the moon's umbra, which was still several miles away, led us to extrapolate a worst-case scenario where these dust devils built in both number and intensity as the cylinder of darkness came nearer. Were these wisps to merge into one another, they would create a fog of sand grains in violent motion. At the very least there was the potential for them to sandblast unprotected cameras and other sensitive equipment. While all of these could be protected, a sand-fog, even if only a few metres high, could be enough to obscure our view of Nature's finest spectacle.

Thankfully this did not happen. Within a few minutes, the dust devils died out, and this episode passed. Presumably the unstable boundary between unalloyed heat and penumbral cool had now passed such that the atmosphere settled into a more stable phase.

The temperature now cooled noticeably as the sun entered its final seconds on view. Although daylight had dimmed to a level more familiar at twilight, the sky above did not appear to be particularly dark. This was a noticeable difference to what I had experienced in Zambia in 2001, not to mention the extreme case of cloudy Cornwall in 1999. The reason, unique in my experience, was that although we were now at the edge of darkness, the rest of the Sahara burned bright—at 'full earth'.

* * * * *

The whiteness of the sand now became a natural screen on which fresh marvels became visible. As the sun was about to disappear we noticed black and white shadows in bands scurrying across the ground. They were visible in the corner of one's eye, and seemed as if millions of tiny ants were scampering across the sand, in successive waves, each just a centimetre across, and separated from the next by a similar width of light sand.

The English Astronomer Royal, George Airy, noticed this strange phenomenon of 'shadow bands', in 1842. He remarked that children ran after the 'strange fluctuations of light' and 'tried to catch it with their hands'.[1] The cause is similar to the patterns of light, known as 'caustics', seen at the bottom of a swimming pool, or the twinkling of stars.

[1] Quoted in M. Littmann, K. Willcox, and F. Espeniak, *Totality: Eclipses of the Sun*, 3rd ed. Oxford University Press, 2009.

Stars twinkle due to turbulence in the atmosphere. A star is so far away that its light comes to us from a single point. If something deflected a light ray slightly, it would not reach your eye and the point would darken. Conversely, if a ray that was on course to miss you were to be diverted into your eye, the image would brighten. Turbulence in the atmosphere causes light rays to be slightly bent as they pass through, and so the number arriving at our eyes fluctuates. This we perceive as 'twinkling'.

The sun is large, and rays come to us from a range of directions, from the top, the middle, and all points around the circle. The subtle deflections of rays from across its entire face cancel one another, on average, and so the sun does not twinkle. However, during an eclipse, the sun is progressively obscured. In the last seconds it is a thin crescent, and in extremis a line of points. Turbulence can now make each of those points twinkle, and as they are precisely aligned, the effects combine in resonance.

Total eclipse thus provides two circumstances that favour the phenomenon. First, the sun has been reduced to a thin slit. Second, the sudden cooling of the atmosphere creates turbulence and also layers with different temperatures. There has been a lot of debate about the actual details, such as under what circumstances shadow bands will occur. It is hard to test the theories however, because the phenomenon is so transient and hard to quantify. Precisely how broad were the bands, what was their relative brightness, for how long did they last? To make such measurements during the excitement of the impending diamond ring is all but impossible. Furthermore, even the existence of these shadow bands has been controversial for more than a century. But they are real: modern video technology recorded them on this occasion for the first time. This was thanks to the white sand, which provided a natural screen on which nature played out this phenomenon.

* * * * *

Diverted by these effects, I looked up just in time to see the diamond ring, and the moon leap into silhouette. This was the moment when I hoped to experience Joshua's illusion, but at that instant I was disturbed by an unnatural human intervention.

A golden rule if you plan to experience darkness at noon is to ensure that you are not near anyone who is likely to shine a light. In Cornwall, for example, cameras flashed as people attempted to take photographs in the gloom. In Libya the assembled watchers were savvy, and had ensured that their flashes were disabled. However, we had not allowed for the bus drivers, who were asleep in their vehicles, the more luxurious of which had air conditioners on. By now this was unnecessary, as the temperature outside was probably lower than the conditioned air within their cabins. The electrical circuits of the buses were active, and with apparent nightfall—for that is what the level of darkness appeared to be, at least if you are a light-sensitive computer—the vehicle's headlights automatically illuminated. 'Turn those lights off', people screamed, though in vain, as English was not the drivers' first language, even had they been awake.

By shielding my eyes from the glare of the headlights, I preserved my night vision, which I had carefully developed over the preceding hour. Even so, the sky did not appear to be as dark as in my previous experience. Far from being like night, it seemed to be illuminated, as if by the floodlights of a distant sports stadium.

This was a real phenomenon, and not due to the headlights having spoiled my vision. For there was a very noticeable further effect on this occasion: the moon was not black as coal.

Outside the immediate umbra of the moon's shadow, the desert was in the full glare of the midday sun. For hundreds of miles in all directions, we were surrounded by beige sand, which acted like a mirror, returning the rays upwards into the atmosphere and beyond. As a full moon can illuminate the night here on earth, so at 'full earth', the Sahara would shine like a lamp, with a black smudge—the Moon's shadow—traversing it. This was why the sky glowed, as if it were a moonlit night rather than a total eclipse. Had there been astronauts on the moon's surface at that moment, they would have found their own night sky far brighter than the silvery glow that we experience at full moon. The moon's surface would have been illuminated by an earth, which appeared as bright as a car's headlights.

And indeed, that is what we could see at mid eclipse, when the moon completely obscured the sun. Instead of a disc of pure blackness, we could discern features of relative greys, as earthshine exposed the mares on the lunar surface. Thus we had the bizarre vision of the black eyes and face of the moon, familiar against a white background when the moon is full, visible against the dark grey of the absent moon, thanks to earthshine.

Every eclipse is indeed unique. This is the only occasion on which I have seen earthshine reflected onto the moon.

* * * * *

With totality over, and the memory of our lengthy trek to reach the site still fresh, everyone was eager to return to the buses and get back to the ship without delay. If we left within an hour there was a chance to be there in time for supper—an important goal, as we had eaten nothing more than sandwiches during the eclipse, to supplement a coffee and croissant at 2 a.m. So we all climbed in, Bill and Cathy joining us in a decent coach for the return. An Arab speaker interpolated with the driver. It seemed to have worked, as within 2 minutes the bus was full, and off we drove—but in the wrong direction.

Instead of returning towards Benghazi we continued south for a mile, and pulled off the highway and onto a track, to be greeted by a surreal sight. 'Eclipse City' was more than a tourist headline, it was a physical entity: a temporary encampment had been built in the desert. There were hundreds of tents, which had been available for us to use as shade; some grander tents, large enough to have hosted a party if anyone had been so inclined; and row upon row of what looked at first to be beach cabins, but turned out to be portaloos. Entrepreneurs had set up stalls to sell T-shirts, which celebrated the eclipse. There were at least three independent teams of television cameras and reporters, who had presumably been there to film reactions during the eclipse. There were also several swarthy, athletic men, who seemed to be more like covert security staff than eclipse chasers. Sporting boots, neatly pressed slacks, and shirts that would have graced Savile Row, their sartorial elegance was hardly de rigeur for the Sahara.

No one had thought to tell any of us about this encampment beforehand, which was presumably why we had spent the eclipse in the wastes of the Sahara. Having bought a T-shirt and used the facilities, we were now even more eager to get back to the ship, but continued to be held up for some reason. Then events became even more bizarre.

We heard a deep throbbing sound, which grew steadily louder. It source was suddenly revealed as a large military helicopter, which appeared over the dunes, circled the camp, and then landed, about a mile away. The swarthy athletes were now revealed to be security men. They began to supervise soldiers, who had appeared as if from nowhere but we realised must have been encamped in the large tents. An orchestrated demonstration of support then took place in the vicinity of the helicopter while television crews filmed as Muammar Gaddafi, or someone made up to look like him, in dark green uniform, red epaulettes on his collar, and the trademark sunshades covering his eyes, stepped from the helicopter onto the sand.

He was too far away for us to hear him clearly, so even if we spoke Arabic it would have been impossible to understand what was going on. But from his expansive motions, waving at the sky and the still partially eclipsed sun, I was reminded of that earlier eclipse in the African continent, Zambia in 2001, where the small boy had asked me if the government had organised the eclipse to make money. Without doubt, Gaddafi had opened Libya to tourists for the eclipse, and was now making the most of this—and why not. But I would not have been surprised if, as his body language and gestures suggested, he was claiming credit as the impresario who had organised the natural phenomenon itself.

* * * * *

The propaganda complete, we were free to go, at last. We re-joined our buses, and made bets as to how long they would take to reach the ship. Bill wisely waited until everyone else had made bets, then he asked which was the most pessimistic forecast, added another half an hour to that, and made his own wager.

We might have reached the ship by suppertime, but Gaddafi's appearance seemed to have unnerved the police. Now the outriders insisted that we go no faster than about 40 miles an hour, and remain in a neat convoy. It was as if our sight of Gaddafi had elevated us to god-like status, whereby we could now only travel through Libya in a formal procession. Drivers of buses further back in the convoy, however, were unaware of what was going on at the front. They started to speed up and to overtake one another. Soon the first buses from the rear reached us at the front and swept past, onwards towards Benghazi.

The police outriders turned on their sirens and flashing lights, as if in pursuit of escaping bank robbers. Within a few minutes we caught up these buses, which had been stopped at the roadside by the police. We now all had to wait until the convoy was neatly reassembled. Only when this had been done to the police chief's satisfaction were we allowed to proceed, in orderly fashion, once more. This saga was repeated on at least two occasions until, like recalcitrant infants, we learned our lesson and were brought into line, in this case, literally so.

At last we reached Benghazi. The sun had long since set, and the city was darker even than the total eclipse. It had been dark too when we left in the pre-dawn of that morning, so I never saw the city by daylight. My memory of that day is that at sunrise I had been already transported magically into the Sahara, where I had watched as the sun was first eclipsed by the moon, was reborn, and eventually set over the western sands of the desert. And that within minutes it was followed by a pencil thin curve: the most beautiful and newest of new moons that I have ever seen.

We approached the port. Given the disordered nature of the day's events, I would not have been surprised to find the gates shut and bolted for the night. But for once we were lucky. We clambered out of the buses, exhausted, and climbed wearily up the gangway of the ship. It was nearly midnight.

We had taken nearly 10 hours. Bill's wager was the nearest, but as even he had underestimated by over an hour we decided the wager was null and void. We were desperate for a drink, and rushed to the bar, only to discover that Libyan law forbade alcohol to be served as we were still in port.

The crew verified that everyone was on board. The dull thud of the engines showed that we were ready to leave, and with relief we began to edge away from the dockside. The ship moved gingerly across the harbour, with a pilot on board, and a customs boat accompanying us. Five miles from shore, the boat drew alongside. The pilot and a Libyan customs officer descended a ladder from our ship to join their colleagues in the customs vessel. As their boat circled around to return to Benghazi, our captain announced that the bar was open.

Only then, as we shared our memories of that remarkable day, did I realise that I had missed my goal of identifying Joshua's phenomenon. At the moment of totality, the shadow bands and vehicle headlights had distracted me. One moment there had been a sliver of sun; the next there was the silhouette of a moon backlit by earthshine. The illusion, if illusion it was, had happened in the blink of an eye. Regrettably, on this occasion, I had blinked. There was still more for me to discover, perhaps next time.

8

The Most Remote Eclipses

Four years later, on a hot July morning, my wife and I were at Los Angeles International airport to meet Cathy and Bill 'it had better be good' Colglazier. They were arriving from Washington DC. That night we would all fly to Tahiti, where, after some relaxation, we would join a ship full of eclipse chasers and head off into the vastness of the South Pacific Ocean. Our goal: to see, on 11 July 2010, what has been described as 'the most remote eclipse in recorded history'.

We had foregone the eclipse of 1 August 2008, total in northern Greenland, Novaya Zemlya, and Mongolia. After the Sahara Desert, a trip across the world for a similar experience in the Gobi Desert wasn't top of my agenda. Also, I missed one of the more unusual eclipse hunts: totality viewed from an aeroplane.

The idea of chasing the moon's shadow in a jet aircraft wasn't new. Years before, on 30 June 1973, a prototype of Concorde was specially adapted to give a group of scientists an extended period of observations of the eclipse as it tracked across Africa. The shadow moves even faster than Concorde, but the plane managed to keep within the spot of totality for more than 74 minutes. The Greenland venture involved a regular commercial jet, which made a round trip from Germany to the vicinity of the North Pole, intercepted the eclipse en route, and extended a 150 seconds length of totality by half a minute.

I learned later that the expedition had mixed success. The plan was to fly across the Norwegian and Greenland Seas for several hours, with the only diversion being a 'sightseeing' overpass of Svalbard. The plane would then meet the onrushing shadow, loop around, and fly within it. The full experience would involve having access to the windows on the left side of the plane, as it would be only from this side that the eclipsed sun would be visible. Some accountant had the bright idea of selling more seats by extending the journey for another couple of hours so that passengers could have a glimpse of the North Pole.

One had to admire the marketing ability of the organisers. They managed to spin positives out of potential negatives, such as the competition for limited viewing of the eclipse through the cabin windows. The advance publicity that I saw hedged the question of how many seats would be sold, and how many would have uninterrupted view of the eclipse. If ever you wanted to take a 12 hour round trip from Dusseldorf to Dusseldorf via the North Pole, this would be the way to have done it, but as for seeing a total eclipse, even extended by half a minute as the plane flew at 500 mph at 37,000 feet, no thanks.

The next opportunity, on 22 July 2009, was a track that crossed Japan and South-east Asia, and lasted a mammoth 6 minutes and a half, which made it one of the longest total eclipses of recent memory. (The maximum possible time for totality is a little over

7 minutes, which will also occur in July, but not until 2186.) We could have watched in Shanghai, but after our experience with the bus headlamps in Libya, I feared that darkness in Shanghai would probably induce a kaleidoscope of coloured neon like Times Square.

We had foregone the 2009 possibility for the more exciting challenge of the South Pacific, which was due on 11 July 2010: the classic gap of a year, less 11 days. In a remote part of the ocean, which would have required some weeks to reach, totality would last more than 5 minutes, but we had made a compromise between inaccessibility and duration, and settled for a ship that would cruise a few hundred miles from Tahiti and the islands of French Polynesia to intercept the eclipse path. We had waited four years since that tortuous Libyan expedition into the Sahara Desert, but Bill and Cathy's arrival at Los Angeles International confirmed: it must have been good.

* * * * *

We children of the 1950s had been brought up with the tales of *Robinson Crusoe, Treasure Island,* and *Mutiny on the Bounty*. How could we not have a romantic vision of the South Seas, where palm trees shade white sands of coral, washed by the calm clear waters of a lagoon. That we were about to visit this idyll at last, made even the cramped confines of the full Air Tahiti plane seem an exciting part of the adventure. Unable to sleep, we flew through the night, over thousands of miles of anonymous Pacific Ocean.

I mused on how, as a child, I had once been taken to a local department store before Christmas to visit Santa Claus. With my parents, I had entered a small room and sat on a long bench while we set off on a 'journey'. Images on a screen gave us the impression that we were travelling from Peterborough, past snow-capped peaks to Santa's kingdom. The film stopped, lights came up, and we exited through another door and, miracle—there was Santa in his grotto. After seeing him and informing him what I wanted for Christmas, while my parents listened, we returned to the store by the simple act of walking five yards round the far corner of this make-believe land. Somehow this failed to destroy the illusion that I had made a magical journey. That is how it can be when you are only five years old.

As we flew through that night with Air Tahiti, I was nearer to 65, yet the magic was still there. I had entered a door at Los Angeles International Airport, and sat down inside a long cigar together with some two hundred other people. We felt our seats dig into our backs, then lift us into the air, after which we were shaken around for several hours before we descended, and landed. There had been no snow-capped peaks along the way, just darkness and the occasional glimpse through the windows of the Milky Way and southern constellations of stars, but when the front door of the plane re-opened, I could have been five years old again, about to enter the magic kingdom.

No Santa in his grotto, but balmy air even though it was still night. As we walked across the concrete between plane and immigration hall, I could smell sweet aromas of flowers, and feel a sense of humid warmth to come.

It was strange that after travelling almost half the way around the world, we had apparently ended up in Europe. French Polynesia was all but an extension of France itself. European passports were waved through, and the euro was the preferred international currency. And on 14 July, Bastille Day, the French national holiday, everything would be closed.

We had landed in Tahiti three days before embarking on our cruise, in order to acclimatise and enjoy the balmy South Pacific. At our hotel we would try to adjust our internal clocks, which would be almost half a day out of synch with home. Dawn was still an hour away as we fell onto a king-sized bed, the sounds of the ocean audible through the open patio door, and fell asleep. I hoped that when I awoke, the dream would still be reality.

* * * * *

Like druids, who gather to greet equinoxes at Stonehenge, I had joined an international cult whose members worship the death and rebirth of the sun at moveable Meccas, about half a dozen times every decade. Ten years earlier, I would have regarded such activity as sign of mental imbalance, if not downright weird. Even so, having made my commitment to the dark side, I was not prepared for how far the spectrum of weirdness extends. It is not solar eclipses as such that create oddballs, at least not as far as I am aware, though I am hardly a neutral witness. Rather, these singular spectacles attract some individuals who are already living in a fantasy parallel universe. These include believers in UFOs, alien abduction, or that they themselves are in communication with aliens. The extreme delusion of someone in the latter case is that aliens are using that particular person as the conduit for them to land, in preparation for the dawn of a new age of enlightenment.

Thus, with hindsight, I should not have been surprised that en route to the twilight zone I met two such individuals. On our first evening there was a cocktail party by the hotel pool, where we found ourselves among a small group of English speakers. Ironic, perhaps, was that most of the people at that poolside party were not there for the eclipse and were even unaware, apparently, that this singular event was about to happen in the vicinity.

I was chatting with one of the few who did know about the eclipse. In the course of the conversation he became aware that I was a physicist. My new acquaintance, a dumpy man with a florid face that looked as if it was far too late for his floppy sunhat to have much purpose, then made his gambit: 'Ah! A physicist. Do you know Professor Hawking?'

The nonchalant folksy way that he dropped this into the conversation sounded as if he was making a point of contact, casting a fly to find a common friend from whom all manner of tales linked to the subject would then lead to a convivial time together. It is, however, the wrong question, as everyone knows Professor Hawking, much as we all know the Queen or David Beckham. What the questioner really wants to know is whether the celebrity knows you.

Answer 'Yes', and you will be in danger of being probed for confidences, which once given will be passed among the salons of Tunbridge Wells, or wherever your inquisitor lives, the moment that they return home. Reply 'No', and you have shown yourself to be less interesting than at first might have appeared. So I hedged: 'Why? Do you?'

It seemed not, although it was possible that the good professor knew of him, as it turned out that my acquaintance had bombarded Hawking with letters and emails, which contained details of 'a unified theory of the universe'. The professor, I learned, had declined to respond—wisely in my opinion—and so had let my acquaintance develop the theory in such a way that any credits would be for him alone. Hawking having declined to respond, it seemed now that I had just won second prize in the queue to learn of these great truths, which, my acquaintance assured me, had hitherto escaped the most brilliant scientific minds.

With fanatical fervour, he prepared to share his discovery: 'Did you know that the wavelength of the gold atom equals the ratio of the diagonals of Orion?' he began. All the words are in the dictionary, but gathered in such a way they left me in the dark. My face must have shown my doubt, because he carried on: 'And this also matches the dimensions of the pyramids.'

I glanced nervously around me, but there was no easy means of escape. Instead, I replied lamely: 'It makes no sense to me.' 'Ah', he leaped in, 'that's because you're not an alien'.

He then leaned in closer, his body language promising that something strictly confidential was about to be passed over, and in a low voice added: 'The aliens have landed. I've cracked their code. The pyramids and Orion, it's all there if you know how to interpret it.'

I hadn't come half way round the world in order to meet a fellow countryman who was, in the jargon of my trade, 'off the real axis'. His confirmation that I am not an alien, gave me an inspiration: I decided to play a trump card. Matching his sense of conspiracy, I too leaned in and shared a confidence in little more than a whisper. 'Look, I have a confession. I am not really a scientist. I'm here in secret and it's important that the staff at the hotel and the crew of the ship don't know the truth about me. As there will be lots of scientists here for the eclipse, I decided to hide in the crowd by pretending that I'm a scientist too. You see, I'm really the travel writer for a national newspaper and have to operate undercover so the crew don't catch on.'

He seemed to accept this, and looked honoured to be party to this deception. I hoped too that he would alert the staff in the hotel and its restaurants that the British analogue of the Guide Michelin's mystery shopper was in town, so that I would receive impeccable service. The other good news was that he was not part of the sea cruise to the total eclipse: he would be watching the eclipse from the hotel, here in Tahiti.

Which must have been a shame, as after travelling 12,000 miles from England to the far side of the globe and Tahiti, he would have seen no more than a 95% partial eclipse. Dramatic, maybe, but: 'so near and yet so far'.

* * * * *

After two days enjoying Tahiti, and with a day still remaining before the cruise, we began to notice the arrival of serious eclipse chasers at the hotel. One seemed to be known to all. This was a tall man, sporting a hat with several eclipse mementoes attached, and a cigarette usually close to hand. This turned out to be none other than Bill Kramer. Bill is one of the foremost calculators and illustrators of eclipse predictions. A lanky American, with a dry sense of humour, who could double for Will Self, he now lives in Jamaica when not travelling to eclipses.

Among Bill's specialities is predicting the duration of the eclipse, its path, and the nature of the diamond ring, the flash of light as the sun shines through valleys on the moon's surface at the start of totality. This is the most beautiful sight in nature, according to Bill—and he has travelled far and wide.

Every solar eclipse starts at sunrise at the beginning of its track. It then crosses the earth's surface from west to east, until totality ends at sunset on the other side of the world, several hours later. As I said earlier, the sun is about 400 times bigger than the moon, and is about 400 times further away. This cosmic coincidence means that the moon can completely obscure the sun if it passes directly between our nearest star and our line of sight, which is the source of the total eclipse. However, the moon is not a perfect sphere, being covered with mountains and valleys. As the moon finally is about to obscure the sun, some mountains cover the sun's disc early whereas valleys are momentarily still allowing the last slivers of sunlight to pass through. The character of the eclipse and the orientation of these lunar valleys change along the route, however. As the moon orbits the earth, its distance from us also changes. In addition, the moon wobbles slightly so that the cross section as perceived from earth varies throughout these hours. So the positions of the last flashes of sunlight through lunar valleys change throughout the eclipse path. Bill uses charts of the moon's topography, and knowledge of its orientation, to compute which valleys will play starring roles, and thereby he predicts the position and time-span of the diamond ring.

He can calculate the duration of totality to a precision of one tenth of a second. This is important for photographers and those who are observing the ghostly solar corona through telescopes during totality. The actual time of start or finish he calculates to about 1 second, as it depends sensitively on your location. 'Anyway, who really cares about that?' he explained. 'If I said we're going to have a total eclipse sometime in the next 15 minutes, who wouldn't wait!'

* * * * *

En route to the eclipse site, which was some 200 miles south and west of Tahiti, we island hopped across French Polynesia. Ever since the 1950s, when I first looked for the exotic South Pacific, which seemed to be the home of solar eclipses, I had wondered if I would ever visit this idyll. The musical, *South Pacific*, made Bali Ha'i sound like paradise on earth. Polynesia was the stuff of dreams. When finally I saw it, there were indeed palm trees leaning over an azure sea, and lagoons filled with coral. Reality, however, wasn't always as the travel brochures advertise.

The outlines of Bora Bora and the lagoons were familiar, probably because I had seen them in scenes from the movie *South Pacific*, and as the backgrounds in countless commercials. In this paradise, however, there are reminders that it is still the real world, where the second law of thermodynamics cannot be overruled. We need energy, and its production generates unavoidable waste. As we drove along the coast road of a beautiful island, and marvelled at the gorgeous villas, which mingled with more humble dwellings, ubiquitous green and grey wheelie-bins lined the road.

Had we visited on any day other than when the garbage trucks were due, these would have been hidden away, to leave a more pristine version of paradise. The Elysian vision would, however, have been temporary at best because round a bend, beyond the houses, we came across a beach that had become an unofficial dumping ground.

At the borders of the most perfect of turquoise lagoons were wrecks of cars, their paint long gone, streaks of rust covering their door panels, their bonnets nothing but brown oxides of decay. Rubber tyres were strewn at random, along with oil drums, bits of reinforced concrete, and the odours of canine faeces. Cars built in Japan, the USA, and France are driven until they break down, whence they are stripped for spares and the carcass abandoned at the roadside on some piece of wasteland. What must have been an idyllic paradise a hundred years ago has become captive to industrialised society. Signs of the international recession had reached paradise, with hotels closed, their windows boarded over. My feelings were very mixed. While we visitors came through on our tour bus, and brought wealth to the islands, their natural beauty was blemished by human intervention. Beyond the villas of the rich and the raw natural beauty, there was underlying poverty and shabbiness.

* * * * *

The traveller who believed that the aliens had landed already, and that they live among us, thankfully remained on land as we set off by ship. If billions can accept that someone in human form can ascend to heaven in a cloud, which is a genuine miracle, then a belief in alien invasion, which at least violates no fundamental axioms of modern science, should perhaps not be dismissed. More extreme are individuals who believe that they are the earthly contact for aliens who are already en route.

If you sincerely believed that the aliens were on their way, and expected to meet you when they land, how on earth, literally, does one make the rendezvous? When the ship left port, en route to the eclipse, one among us believed that they had the answer.

Given the magical and mystical culture surrounding a solar eclipse, there is a certain logic that the earthly arrival of aliens from other worlds will coincide with such an event. That, at least, was what one member of our party in the Pacific Ocean seemed to have concluded.

The first clues of something odd came at breakfast on day one. There were about a hundred passengers in total, all there because the ship planned to intercept the solar eclipse. An azure sky reflected in the Pacific Ocean, which gleamed with hues of turquoise and darker blues mingled with the sparkles of a million diamonds as sunlight scattered off the gentle waves. The air was warm and dry, fresh without a hint of pollution, and sensuous on the skin. The attire: shirts and shorts for the men with brimmed hats and bared shoulders for the women, sunshades and cream for all. Or, rather, that was so for almost all.

One man wore a shabby sports jacket and tie, with canvas trousers, and hair unkempt. Sandals and bare feet were his only concession to the occasion, and he looked more like an English schoolmaster on a weekend retreat than someone on a cruise in the South Pacific. This appearance was completed by the fact that instead of a backpack, which was the normal appendage of the group when carrying their belongings on deck, he carried a briefcase, similar to the Gladstone bag favoured in the nineteenth century.

The crew had arranged tables on the open deck in the morning sunshine, prepared for breakfast with places around each set like a bridge four. He carefully arranged his bowl of fruit and organic cereal at one place on the square table top, with a book already opened to warn off any attempt at social interaction. He completed his defences with three neat piles of books, one at each of the three empty places on his table. Each pile was positioned accurately at the midpoint of its placemat, the individual books aligned vertically with a bricklayer's precision. Four glasses of orange juice were placed at each of the locations, their volumes obsessively measured to be identical.

Satisfied that the symmetry was perfect, he leant down to his briefcase and removed a pair of headphones. These were not the miniature earpieces that are a modern fashion item, but a huge pair of mufflers, which looked like aviator's communication equipment. Only now, suitably dressed for breakfast, did he begin to eat.

By now he was the centre of attention, but his desire for privacy was a success. As people finished their breakfasts, they retired to the pool, or to deckchairs, to enjoy the presence of paradise, and left our strange companion to his own introspection.

Later that morning, however, he appeared once more, on the upper deck. Whereas we were in bathing gear, or shorts and casual shirts, he was dressed as before, his headphones now draped around his neck. Like a magician, from his bag he produced various *objets d'art*: a wooden sextant, a pair of dividers, a protractor, and an old-fashioned slide-rule. From the inner pocket of his jacket he took a pencil and a notebook. He aligned the sun in his sextant, and performed calculations with his slide-rule. He then entered these data carefully into his pocket ledger. This ritual was repeated every morning.

Attempts to communicate with him received monosyllabic answers. He responded to requests for explanation of how his tools worked but his replies made little sense. All we could disentangle was that the world was in peril and that he expected someone to collect him. This was to happen during the voyage, but how, where, and when remained a mystery to me. The rituals with his tools were apparently designed to ensure the rendezvous succeeded.

This was repeated up to the day of the eclipse, at which the pattern changed. While most on board prepared their cameras, or selected their deck chairs for the show, he came up to the prow of the ship, together with three suitcases. Several among us had brought a set of specialist equipment to observe the eclipse, record the event with telescopes, cameras with special filters, even a collection of devices with an all-round view, mutually controlled by a laptop computer. His suitcases, however, contained none of these. Instead they carried his essential clothes and personal treasures. The explanation, we conjectured, was that if aliens were to appear during the eclipse, and evacuate you, then it would be sound planning to have your prize possessions to hand. We were assured that members of the crew were also keeping a careful watch, lest he jump overboard. Not just the passengers had noticed his strange behaviour.

Meanwhile, other members of our party opened up their bags, which contained more conventional equipment. Tripods were aligned along one side of the ship, which, if all went to plan, would be oriented with the best aspect for the sun.

* * * * *

An eclipse at sea has its own special features. A ship can be bad for some professional photographers, who need a very stable base for their telescopic lenses, but is good in cloudy weather, as the vessel can manoeuvre to find clearings in the overcast. In extremis, if the weather turns really bad, one can change course and look for a window in the storm.

We found an area on the eclipse track where the sky was clear, save for cumulus clouds, which looked like puffy pieces of cotton wool. In England this would herald a perfect day. In the South Pacific on eclipse day, however, it created a problem. The eclipse would be relatively far in the west, at an angle of only about 20° above the horizon, and in such a case an effect of geometry, of perspective, gets in the way. There were large gaps between the clouds directly above us, which would have been good news if the sun had been high in the sky. Viewed from afar, however, the line of sight makes even occasional low-lying clouds obscure large amounts of the distant sky. Remote clouds appear to layer up like bricks (see Figure 8A).

You can check this for yourself. On a sunny day, when there are only a few clouds in your vicinity, take a look towards the horizon. You will have the impression that the weather is less good over there. Even though you are in a sunny clearing, the rest of the sky appears to be clouded over. So the chance of finding a

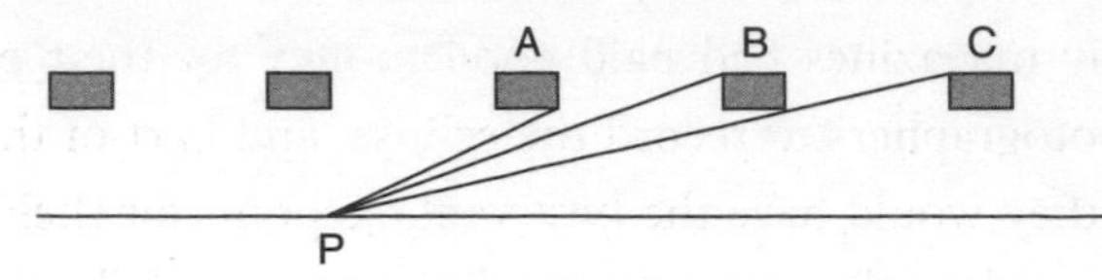

Figure 8A. The brick effect. In this illustration, the black boxes denote clouds and P a person looking at the sky. Three quarters of the sky is clear of clouds. The lines show the lines of sight to the base and top of nearby clouds. The gap between clouds A and B is already only marginal, and as B overlaps C, the sky appears to be completely cloud covered from B onwards.

low angle line of sight through the clouds to the low-lying sun might be difficult. It began to appear that we might have to do some tricky manoeuvres in order to get a clear view of the eclipsed sun.

The ship travelled along the eclipse path for a few miles as the leader of the expedition assessed the movements of the clouds. During the hour's build-up of partial phases, the sun alternately disappeared and reappeared behind clouds. As totality approached, however, the clouds seemed to have become more numerous. Even with most of the sun obscured, and twilight descending, it was still possible to see thin shafts of sunlight between the clouds, which gave clues as to where we should head for a better view. As darkness began to fall, sunlight would twinkle on the surface of distant waves even while the sun itself remained obscure to us.

Totality was now only minutes away, and the ship was too large for rapid changes of tack. Then at almost the last moment a clearing developed. However, to use it to best advantage, the captain would have to turn the ship through 90°. Some international astronomy magazines had paid good money for their professional photographers to record the eclipse, and part of the deal was that they would have the best vantage point for their cameras. The result of this last minute manoeuvre would be that the experts would no longer be best placed on the ship. It was too late for them to relocate efficiently, but as the alternative would have meant no pictures anyway, the ship turned to port, and the rapidly shrinking sun miraculously appeared in the middle of a patch of clear sky.

The uninterrupted view to the horizon in all directions gave a dramatic panorama of the moon's shadow. We saw it first over the far horizon. The sea was dark green, with faint twinkles from the last rays of the dying sun. Earlier, the horizon had been a clear circle against the darkening sky, but now it disappeared, as sea and sky seemed to merge into one. From beyond the horizon, dark grey welled upwards and outwards to fill the distant sky. It was so dramatic, as on this occasion the sun itself was low in the western sky, and so we were looking at it and the approaching night at the same time. The moon's shadow rushed over us at supersonic speed. Once again we marvelled at the wonderful sight.

The blue of the sea was scattered into the sky. The moon's silhouette took on a dark green hue, its origin like the earthshine that we had experienced in Libya, but this time with the reflected colours of the ocean. As the moon finally obscured the sun, their low elevation in the sky obscured any sense of their motion, so I was unable to tell whether the moon had moved back, forwards, or anywhere at all. Its shape just popped into existence where, a moment earlier, the sun had been. The moon looked like a dark button that had been sown onto a turquoise quilt while my attention had been elsewhere.

No alien spacecraft appeared. The bags of the intended traveller returned below deck, along with their owner. He passed the rest of the trip, apparently unperturbed by the failed rendezvous.

For the rest of us, for whom the purpose had been to watch the eclipse, the positioning of the ship had worked perfectly. The eclipse was in the clear until the end of totality. A cloud hovered just below the drama, and for the first time I saw one with a genuine silver lining, thanks to the ephemeral light from the solar corona. As totality ended, and daylight suddenly returned, there was a brief shadow as the cloud finally moved and obscured the sun. We had indeed been lucky, thanks to the skill of our captain.

This was my first eclipse where the sun had been low in the sky. It had enabled me to watch the moon's dark wall approach as background to the main event. As for understanding Joshua's illusion, however, it helped not at all. For Joshua had seen an eclipse high in the sky. At least I was now aware that a solar eclipse brings with it unusual optical illusions, which can depend on location, elevation, and, of course, weather. The good news was that I could have another chance of a South Pacific eclipse in just a couple of years, in November 2012.

* * * * *

Through the Wormhole

In 2010 the eclipse had been out of Tahiti; this time—November 2012—we would go further, and cross the dateline to Fiji.

With a 12-hour time difference between the eclipse site and England, I had acclimatised by spending two days en route, on the west coast of California in Venice Beach.

Drifters and dreamers from across the continental United States had travelled west in search of Xanadu. They had continued all the way until they reached California. There being no landmass further west, they deposited their sleeping bags on the waterfront of Venice Beach, opened their guitar cases, and played for dimes. Time seemed to have been suspended, as hippies paraded the waterfront, their hair and dress identical to what had been de rigeur when I had seen that orange moon back in 1971. The hippies were by now grandparents, but other than the obvious signs of ageing, were much the same. It was as if the 40 years since the night of that lunar eclipse had never happened.

Two decades before that, when my interest in eclipses began, I had grown up with the impression that California was where land finally gives way to the oceans, and that the world itself ends just left of Tahiti and Hawaii. The map on the wall of St Mark's County Primary School in Peterborough was typical of many in Britain at the time: Britain sat proud and pink in the centre, with countries of various hues spreading to right and left reaching singularities beyond Japan and the west coast of America. Since then I had learnt enough science to know that not everybody considers Peterborough to be the centre of the universe. Even so, it had been a shock to discover a map on the wall of a bank in the Tahitian capital, Papeete, in 2010 showing Polynesia in the middle, with Peterborough hanging precariously on the edge.

Reassuring myself that this singularity in space was not real, merely an artefact of the coordinate system, two years later I sat in an Air Fiji plane at Los Angeles International, in the late afternoon of 7 November 2012. The flight would carry on to New Zealand, where it would arrive at their breakfast time; the intermediate stop in Fiji was timed for 2 a.m. by the Fijian clock. As we departed Los Angeles, midnight was somewhere over Europe and rushing westwards faster than the plane.

We headed away from California, across the Pacific, with nothing to see but the sea. The vastness of the Pacific Ocean becomes apparent when the only difference in the view from the window hours later is daylight turning to dusk over that interminable sea. Midnight caught us as we crossed the International Date Line. Instead of putting our watches back 1 hour, as had been the case on crossing each time zone during the westward trip so far, here we jumped forward 23 hours as we left the left-hand edge of the map and reappeared on the right. If we had we been superposed onto my childhood Mercator map of the world, then in that instant we passengers would have been temporarily disconnected from the pilot who was 24 hours in front of us, and heading forwards.

The plane stayed in one piece through this strange manoeuvre. So far, so good, but now came the conundrum. As a result of the midnight hour, Air Fiji had swept us not into tomorrow but straight into the day after. In England, people were having elevenses on 8 November. Midnight of 8/9 November was rushing towards them from the Far East where it had already overtaken me. My digital watch refused to admit I had jumped from 7 November into the ninth and been expunged from the eighth.

My mind numbed with accumulated jet lag, and confused, I attached my seat-belt as we lost altitude and skimmed the waves before touching down on dry land. A pleasant warm humidity greeted us as we set foot on Fijian soil at 2 a.m. to be greeted by dancers, wearing garlands of flowers, and strumming ukuleles. I had no idea what day of the week it was and hoped that the ship's captain would get the coordinates right for the total eclipse. It would occur in the middle of the day, that much was clear. The problem would be—which day? The eclipse path crossed the date line, and all but a small part of the event would be on the 14th. However for us, the rendezvous would be on the 15th. Hopefully someone had acclimatised long enough to have cleared their head and double-checked the location. With a horrible fear that we might turn up a day late, I collapsed into a fitful sleep.

* * * * *

Among our party once more was Bill Kramer, the veteran of some 15 total eclipses whom I had first met in Tahiti. Bill's love affair with eclipses began in 1972, when his father—who had never seen one himself—took Bill along to see one at sea off the coast of Canada. They were so overwhelmed that the next year Dad took Bill with him to Africa to see one there. 'I dedicate every one to my father', Bill told me, as he anticipated this latest eclipse with all the excitement of 40 years' experience.

According to Bill, the eclipse path was due to cross the northern tip of Australia and then sweep over the Pacific Ocean, never again touching land. Putting our trust in science, and the ability of people like Bill to compute, our ship had travelled for two days and three nights out of Fiji in the assurance that an eclipse would occur just after 8 a.m. local time at the appointed location.

It was during those days travelling across the blue Pacific, that I began to appreciate its vastness. During this time we saw no land, not even a tiny desert island; we seemed to be at the centre of an infinite empty ocean. No vapour trails spoiled the blue symmetry of the sky. No great circles between cities in that far away inhabited world came anywhere near our isolation.

As the sun rose on that third morning, the view from our deck was of water extending to the horizon in every direction, seemingly no different than the scores of barren liquid horizons that we had traversed along the way. There was a clue that this place in the watery desert was special however. Having seen nothing but sea and sky for three days, about a mile from us was a yacht, bobbing in the waves. Either this was a remarkable coincidence, or we were not alone in trusting Bill.

We learned later that a cruise liner, with over a thousand passengers, had set off from Sydney, Australia hoping to include the eclipse in its itinerary, but was delayed and never made the rendezvous. They saw about 98% of the sun obscured—that is a 98% partial eclipse, interesting certainly, but not totality. When it comes to experiencing a total eclipse 'nearly is not enough'. Totality is something utterly different, and of that they saw nothing. What sights they missed.

* * * * *

I thought back to the young lad in Zambia who, in 2001, had told me that if the predicted eclipse happened: 'then I will believe in science'. His street-smart intelligence could have taken him to college in Lusaka, and by 2012 he would have been old enough to graduate. I hoped that the faithful appearance of that eclipse had inspired him to 'believe in science'. I daydreamed that he might be teaching youngsters in some Zambian school where on the morrow, like a latter day Mr Laxton, he would use the news media and television images of today's total eclipse to inspire the next generation in Africa.

And now, 11 years later, trusting in science, I am on a ship in the anonymity of the South Pacific, hundreds of miles from any land. About an hour before the main event was due, once again someone shouted the mantra: 'first contact', the astronomer-speak for 'the moon has begun to cross the sun'. Through binoculars, suitably protected against the glare, a small nick could be seen disturbing that perfect circle, and it was growing, eating into the solar disc before our eyes. And, once again, I felt humbled by the predictive power of science.

Watching the moon eat into the sun is like watching paint dry. Or it is for about an hour, until in the final minutes before totality, a host of unusual phenomena began to assault the senses.

At first the experience was much as before: as the moon began to slide across the face of the sun, dusk began to fall. But out on the ocean, and on this occasion with the sun high in the sky, and no clouds to worry about, it was a strange twilight. In Zambia I had seen turtle doves begin to fly low across the trees and vultures coming in to roost circling lower and lower, like at normal sunset, except that at the onset of totality, darkness was so sudden that the vultures had landed in the dark. For us humans also it was strange: the light got dimmer but the shadows didn't lengthen. Here at sea the only apparent animal life consisted of the expectant humans, gazing in wonder as the crescent remnant of the sun got thinner and thinner.

During a partial eclipse, you never see the moon. As the stars are there during the day, but the sky too bright to see them, so too is the moon invisible relative to the sun: the vanishing sun is the only clue to the moon's presence. As the sun evolves from a circular disc, to a solid C, and even to a mere parenthesis, it is too bright for the eye to distinguish the moon from background sky. The impression is that the sun is disappearing.

Where before there was a piece of the sun, now there is sky. This remains the case even when a mere sliver of the sun survives. Only as the moon finally obscures the sun is the main actor revealed.

At the moment of totality, the sun is suddenly dim enough for the moon to show, in silhouette.

The illusion is that the moon has suddenly appeared, as from nowhere. The vision is of the sun giving birth to the moon. As the eye is very sensitive to motion, even the slightest change being noticeable, it is aware of the passage of the moon in the last seconds as it completes the obscuration. This trifling movement is what gives the impression of action, of birth, of the moon bursting from the sun and eliminating all light from the heavens as it does so.

Now the light is low enough that the solar corona appears, filling the sky around the central action, but interrupted by a circular jet disc—the moon. The impression of metamorphosis is like some supernatural sleight of hand. From nowhere, a moon has been created. All this while Baily's beads run around the rim, the diamond ring flashes, and the fine alignment of the rims of moon and sun give the viewer an acute awareness of motion. And in this moment of marvel, did the new-born moon appear to move in reverse? I thought that it might have, and could convince myself that it did. Of one thing I was certain. The instant of totality brings with it remarkable illusions—the birth of one heavenly body as another disappears from view, the last glimmers of rainbow light transformed momentarily to pink and then to a ghastly pallor, and a strange sense of movement, which itself is not quite natural.

There appeared to be a visual illusion known as 'chronostasis', of time suspended, as when a moving second hand on a clock-face can momentarily appear to stop.[1] In the brief moment of lunar birth, I seemed to see the black disc recoil, but hardly had I time to register this than totality had overtaken my senses. I replayed it in my mind's eye and convinced myself that I had seen Joshua's illusion. Was it so, or had I created it after the event? It was by now too late to know, but at last I had something definite to watch for—next time. In the meantime, there were still 3 minutes in the moon's shadow to experience in the middle of the ocean.

* * * * *

Bill Kramer mused that you take from a total eclipse what you bring to it. A spiritual person will see this 3 minutes of ecstatic wonder as confirming the infinite power of the creator, some deeply religious observers even having visions of iconic images in the shimmering corona surrounding the black hole in the sky. Others marvel at the ability of science to predict where and when this singular event will occur.

If—when—you experience a total eclipse, don't spend your time looking through a camera. Make it an *experience*. Watch it, certainly, but make a video recording, and, most important, turn

[1] For more about the illusion of chronostasis see https://computervisionblog.wordpress.com/2013/06/26/visual-illusion-saccadic-masking/.

up the sound. I took Bill Kramer's advice and was astonished by what I heard when I played it back, hours later.

The sound on the recording of the eclipse revealed unexpected delights: people gasping and screaming as if partaking of a mass orgy. Highly educated people, who in normal times are eloquent, broke into a form of Jamaican patois that shouldn't be repeated in polite society. Bill later described the experience as 'like going to a Grateful Dead concert but without the drugs'.

Three minutes later—at least, that's what my watch said, but as always the intense experience seemed to have lasted but a few seconds—a second diamond ring flashed as totality ended. Little red flickers—Baily's beads—could be seen running around the limb of the moon as gaseous prominences on the sun's surface were momentarily visible. Daylight returned with a rush. And a booby flew over our heads. Unknown to us all, there had been birds roosting on the ship, and they had awoken as from a catatonic sleep. Life returned to normal, Bill Kramer dedicated one more eclipse to his father, and plans to see the next eclipse began.

Inspired by Bill's father, I decided to take my children and grandchildren to see a total eclipse. What more inspiring legacy could there be? Now let's see; Africa 2013, Faroe Islands 2015, USA 2017. . . .

9

Atlantic Adventure

I decided that the USA in 2017 would be the time to introduce my progeny to a total eclipse. My elder grandson would then be the same age as I had been when, in 1954, a solar eclipse had inspired my interest in science. Before that, I would have two chances to test my theory about Joshua's reversing moon. I wanted an eclipse near the middle of its path, when it would be high in the sky, in order to have the best chance. The first convenient opportunity would be on 3 November 2013.

That eclipse would be a rare 'hybrid' eclipse, which is annular for some parts of its track, and total in the rest. It happens when the tip of the moon's umbral shadow reaches the earth's surface at some places, but falls short of the ground along other sections of the path. This subtle phenomenon touches on the very existence of total eclipses.

I said earlier that total eclipses are the result of a cosmic coincidence. It has not always been thus, and will not remain so in the far future. This is because the moon is slowly receding from us, at a rate similar to the growth of fingernails: about 4 cm a year.

About a billion years from now, the moon will be too far away to fill the solar disc, and the last total eclipse will have occurred. So we live at a fortunate epoch! Even now, the elliptical path of the moon around the earth makes the amount of obscuration different from one eclipse to the next. When the moon is closest to the earth, known as 'perigee', its diameter appears to be about 15% greater than at 'apogee': furthest. If perfect alignment takes place at apogee, the moon appears to be smaller than the sun, in which case an annular eclipse occurs, where at mid eclipse a bright ring of the solar disc remains visible.

During an annular eclipse, the tip of the moon's umbral shadow is tantalisingly above the earth's surface. The curvature of our planet's surface can bring some parts within reach of the umbra, so that it touches down and makes a total eclipse. In this case one has the 'hybrid' phenomenon. The source of this is the miracle of perspective whereby the moon's apparent size changes as it rises from the eastern horizon, moves overhead, and then falls away to set in the west. The reason is that the moon is nearer to us when overhead than when near the horizon.

If this sounds surprising, consider for a moment that when compared to the breadth of the earth, the moon is not far away. Its mean distance is about 384,000 km, which is only 30 times more than the earth's diameter of some 12,000 km.

At moonrise, the moon is low on the horizon and an earth's radius further away from us than when directly overhead (see Figure 9A). This extra remoteness is only three per cent, but this is a significant amount, sufficient to determine whether the moon might obscure the whole of the sun, or leave an annulus visible. This effect makes the moon appear slightly smaller when

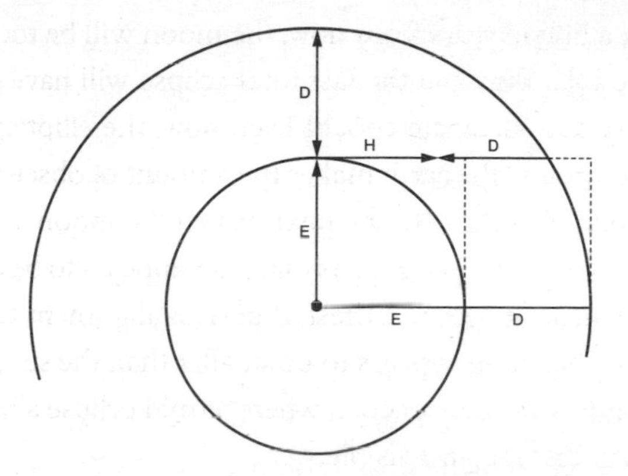

Figure 9A. The moon is nearest when overhead. When overhead, the moon is a distance D away. When seen on the horizon, it is distance D + H away. The drawing is not to scale. In reality, D is roughly 30 times larger than the earth's radius, E. The sun is so far away that this phenomenon is negligible for it. So the moon's ability to make a total eclipse as against an annular eclipse is greater when the alignment is high in the sky.

grazing the horizon than when viewed overhead (the appearance of a large moon near the horizon is an optical illusion due to the presence of foreground objects in the line of sight: the sun also appears larger at rise or set than during high noon). Hence there is the potential for an annular eclipse at the start of the eclipse path to convert into a total eclipse in the middle, before reverting to annular as it heads towards the far horizon at the end of its path.

Thus the earth's curvature can determine whether you see totality or an annular eclipse. The moon's elliptical orbit might also bring it towards the earth during the critical hours, so that what began as annular turns into a total eclipse, but the geometry of the earth's curvature is the more usual cause.

In 2013, at the start of the eclipse path, in the western Atlantic near the coast of Florida, the eclipse was annular: the moon was marginally too far from the earth to obscure the entire solar disc. Within the first 15 seconds of its path, however, the moon's situation had subtly altered. Now slightly larger, it would obscure the sun completely: the eclipse became total. This would be the state of affairs for the remainder of its path, as it continued across the Atlantic, south of Ghana and Nigeria, before making landfall in Gabon, and then onwards across equatorial Africa, via the Congo towards its end, at nightfall, in Ethiopia.

* * * * *

The challenge was where to see the eclipse. At the start of its path, near North America, the sun would be low in the sky and totality would last for only a few seconds. The shadow would pass about 300 miles south of Cape Verde Islands, with an eclipse of just over a minute, its duration being a maximum of 1 minute 40 seconds when south of Liberia. The sun would be high in the sky by this stage. As the shadow crossed Africa, the time reduced to just 1 second at sunset.

The best chance for good weather would be near the end of the track, in Kenya and Ethiopia. However, this benefit would involve a high price: totality would last for just a few seconds.

Equatorial West Africa and the tropics of the Atlantic are in the 'Inter-tropical Convergence Zone', where winds from the northern and southern hemispheres meet. Here, humid tropical air is squeezed upwards, which triggers frequent violent downpours and thunderstorms.

Although this is true on average throughout the region, the areas of storm and clear interludes form a moving patchwork. For a specific location at the moment of totality you might be lucky or, instead, drenched. Half an hour later, the situation could be reversed. The best strategy in such a case is to use the mobility of a ship, as my experiences in the South Pacific had already shown.

The best chance of a clear view looked likely to be off the coast of Africa, where dry air flows off the continent and reduces the likelihood of clouds. The Cape Verde islands offered an interesting location to explore in its own right, and a single day would be sufficient to get from there to the eclipse track.

The group of eclipse chasers, who had charted ships in the Pacific, now adopted the same strategy in the Atlantic. We would fly to Casablanca, spend a week touring Morocco, and then cruise down the west coast of Africa towards Cape Verde, with a stop at Western Sahara.

* * * * *

The beautiful symmetries of the Kairaouine Mosque in Fez, Morocco, would give even the most innumerate a pleasure in mathematics. Geometric constructions, designed with the lengths of their sides controlled by the square roots of rational numbers, intermingled with reflections and occasional subtle and profound asymmetries, all tease the mind. Or you can just enjoy their raw artistic beauty.

Five times a day, the faithful are called to prayer as 'Allahu akbar' sounds from this and the many other mosques in this vast labyrinthine city. No guidebook had prepared me, however, for a remarkable phenomenon that happens when the call to prayer takes place at sunset. This was but one of several examples of how the practice of Islam and the astronomical clockwork of the heavens are entwined.

For followers of Islam, the first sight of the new moon with the naked eye defines the start of the month. So a solar eclipse has a special place, as its termination heralds the wait for the new moon. Only when the line of sight of the moon has separated completely from that of the sun, and the earth has rotated so that the sun has sunk in the west are conditions likely to be ripe for the sliver of the new moon to be visible.

For us the eclipse was still a few days away, and thus it was nearing the end of the Islamic month. Every day of every month, the motion of earth, sun, and moon determine the Islamic traditions, such as the times for prayer: dawn until sunrise, midday—when the sun is at its zenith, afternoon—which is determined by a complicated formula relating the length of an object's shadow to its real length, and sunset.

It was early evening, with sunset imminent, as I stood on a hotel balcony in the hills above Fez in Morocco. Ancient Fez and its modern suburbs were spread out below me, straggling over some ten miles. As the sun sank in the west, I waited to hear the Maghrib, the call to prayer at sunset, when the minarets for a city of over a million people would call 'Allahu akbar'.

What was utterly unexpected, and took my breath away, was the remarkable beauty of what happened next: in modern Muslim Morocco, you can hear the planet's rotation.

At the equator, the earth spins its circumference of some 24,000 miles in 24 hours. At the latitude of Fez the surface speed is less, but still equates to about a mile every 3 seconds. So the sun sets in Fez's eastern suburbs up to half a minute earlier than in its westernmost extremities.

In ancient times the call to prayer was made when sunset was seen, by eye. Today it is computerised. The time of sunset is known, and the start of Maghrib, to the second. Thus it was that I suddenly heard 'Allahu akbar' sound from a minaret in the east, its decibel levels high enough to carry for miles in the still evening air, up from the city into the surrounding hills. Almost immediately another began, then another. There was a strange cacophony as the mosques of the city began their calls, not with some random scatter, but in a wave. From my left—the east—to the city itself in front of me, and then on to the right, the minarets began their calls in sequence, like some stereophonic sound system in a gargantuan three-dimensional theatre.

'Allahu akbar' spread across the land at the speed of sunset. Within half a minute night had fallen everywhere. This daily eclipse of our sun, by our earth, was celebrated in sound.

* * * * *

Western Sahara is a country that I hardly knew existed, let alone ever expected to visit. Even its inhabitants are uncertain to which nation they belong. Most of the area, on the west coast of Africa, adjacent to Mauritania, is administered by Morocco. However, there is a thin strip known as the Sahrawi Arab Democratic Republic (SADR). In my experience, the inclusion of 'Democratic' in a country's title affirms that it is anything but. Morocco treats this strip as a buffer zone, whereas the SADR government considers the Moroccan area of the country to be occupied territory. Hence the ubiquitous presence of many security vans, painted white, and emblazoned in black with the solid capital letters: UN.

Like our experience in Libya, eclipses lead tourists to places where they do not normally go.

I once had a stopover in Cape Verde in 1986. It had been at night, on a South African Airways flight from London to Johannesburg. In that era, the landing strip and corrugated shanty buildings hosted just two destinations: South African airways flights, which were excluded from flying over most African states during the era of apartheid, and Aeroflot, en route between Moscow and Cuba. Crickets had hummed in the humid night air, while the only sign of humanity was a man, chewing a cheroot, who watched us as if in a dream. It felt as if I had stepped into the middle of a novel by Graham Greene.

A quarter of a century later, all had changed. Our bus made a long detour around the edge of a busy international airport, host to charter flights, which bring tourists from Europe. The Cape Verde islands, which are a protectorate of Portugal within the Tropic of Cancer, could become the Atlantic Ocean analogue of the Pacific's Hawaii. The raw beauty of volcanic Fogo would be worth a visit in its own right, as would the greenery, plantations, and farming villages of São Nicolau or the soaring peaks and deep valleys of Brava.

At 17.00 on Friday, 1 November, we raised anchor at Fogo, and headed south towards the eclipse path. Our journey, planned to be 297 nautical miles, was scheduled to take all of the following day, with arrival by the early morning of 3 November. And if the weather had been fine, that is probably what would have happened. However, as I discovered at my first eclipse in Cornwall, weather can change, and when an eclipse is due, this can turn out to be for the worse.

* * * * *

According to the information coming to the ship from a meteorologist with a supercomputer in the USA, there were significant amounts of cloud with the possibility of a hole, but not where we originally planned. Landon Noll, a computer scientist famed for his work with prime numbers, was in our party and sat in the lounge, with his laptop, making his own meteorological computations.

Landon's computer codes calculated a weather forecast in real time. This could give a first assessment, but enabled decisions to be made on possible destinations. These could next be fed to the USA where a meteorologist improved the precision of these predictions on dedicated supercomputers. This took time, so we had precision data only at sporadic intervals. This all confirmed what Landon Noll suspected: cloud cover would build at our intended site and so we needed to change our plans. Two possible solutions showed up.

One was to head east and south, in the region south of Sierra Leone. This would introduce a new hazard, however: pirates. Whereas in the distant past, a solar eclipse could have been a powerful weapon to invoke as defence, taking inspiration from Columbus for example, modern pirates are logged in to the internet and fully aware of what's happening. Indeed, pirates were probably already actively looking out for the possibility of tourists coming into the eclipse path, within reach of the coast. We duly decided to adopt plan B: to head west, deeper into the midst of the Atlantic.

The new location would reduce the amount of time for totality, but at least give a chance of seeing it at all. The ship steered its new course late in the evening, and with good speed should arrive at the new eclipse venue by dawn.

During the night I heard the sound of a change in the engine. The next morning, when I opened the curtains of my cabin porthole, I was surprised to discover that we were still on the move. The sky was slightly overcast, with more worrying banks of cloud visible in the distance. By the central stairwell of the ship there was a chart to show our progress. It revealed that we had turned about towards the mid-Atlantic, travelled for about 50 miles and then zigzagged up and down across the path. The reason was that the weather, although better than where we had originally been headed, was variable.

Weather updates from North America suggested that there would be some clearing by the time of the eclipse. The strategy was to explore the area and then, hopefully, find the best chance for a clear view. Mid-morning the ship stopped. There were lots of clouds, with some gaps, but after all our hard work everything looked now to be a matter of chance. I told some of the glummer faces that an eclipse under cloud has its own interesting character, but no one seemed convinced.

An hour before first contact, the ship started its engines and began to manoeuvre again. There were some thicker clouds at the horizon, and the breeze was pushing them in our direction, but the weather news suggested that there was also some thinning in the clouds over a large area. We found a hole in the cloud cover. The captain positioned the ship so that we could move along the eclipse path to keep in line with this window of opportunity.

Our luck held for about an hour, by which time the eclipse had begun, and the sun thinned to a small arc. The thick clouds were no more, thankfully, but now high-level cirrostratus had set itself in place. We would see the eclipse through a tenuous overcast, like watching Salome's dance but with half of her seven veils still in place.

The gauze of cloud removed just enough of the glare of the dying crescent, while thin enough to allow clear sight of the birth of the corona and the eruption of the diamond ring. This was the first time that I was able to watch the climax of the eclipse with my naked eye, unguarded.

In Fiji I had seen the Joshua illusion—possibly. What I wanted to watch carefully this time, to verify if I was right, was the birth of the black moon. The illusion seemed to be the mind's attempt to make a logical interpretation of a black solid instantly appearing, as if from nothing, when it replaced what a moment earlier had been perceived as part of the background sky.

The sequence now described lasts for less than a second, the crucial parts perhaps a mere one tenth of that. For as long as the shining crescent of the daytime sun is on show, the eye has no awareness of the moon's presence in the drama. As the constellations of stars are rendered invisible in daylight, so is the moon also. This remains the case until the final milliseconds of its transit. Until that moment, the illusion is of the golden sun being cut ever finer within an unchanging background sky, rather than of a lunar disc moving across a bright circular orb.

What happens next could be a project for a neuroscientist to dissect.[1] When events happen rapidly at short time intervals, the mind plays tricks. You can try this for yourself.

Look at your eyes in a mirror. Focus first on the right eye, then the left, and then glance back to the right. Do this rapidly. Your eyes take some tens of milliseconds to move from one focal position to the other, but you never see them move. The gaps in time, when your eyes are in motion, have somehow disappeared. Your brain seems not to care about these gaps in visual input. When the eye moves rapidly between two fixation points—known as a saccade—the mind plays tricks with the perception of time and even the sequence or duration of events.

The events at the onset of totality are a classic example of this illusion. At one moment your eyes are focussed on the fast disappearing sun, at about 8 o'clock say. The next a black solid circle has appeared, as from nowhere. And as you're reacting to this, the diamond ring flashes, and a wraithlike remnant of the original sun now encloses the disc, from 8 o'clock where the last drop had been, to two, six, and all parts between. The focus of attention momentarily moves back and forth. The thin veil of cloud had, on this occasion, enabled me to watch every frame in this celestial video. The sequence of disconnected events, none of which has any point of reference in normal experience, is transformed in one's mind into a coherent narrative.

Where previously there was a sense of smooth continuous motion, these separate quanta, which help define the birth of the moon in silhouette, giving a momentary sense of time suspended,

[1] See D.M. Eagleman, Human time perception and its illusions, http://www.ncbi.nlm.nih.gov/pmc/articles/PMC2866156/.

as one's centre of attention moves back from the crescent sun by about a solar diameter. Your perspective moves backwards; the perception of movement on the celestial stage reacts likewise. This is for but a brief moment. On top of this, however, there was a further sense of motion thanks to the clouds. As these drift across the field of view, they provide your only sense of 'vertical'. Like the onset of nausea at sea, when the mind interprets the moving walls and deck of a pitching ship as the true indicators of 'upright', so the clouds are perceived as the link to the matrix of your environment.

In normal circumstances, as one interprets the clouds as static, this can induce a sense of vertigo. In the excitement of the eclipse, it gave an impression that the sun and moon were drifting backwards. There is no hint in Joshua whether the day was cloudy, and if so whether that added to the illusion. But after six eclipses, spanning a dozen years, I was satisfied that the 'miracle' of Joshua probably is an eloquent testimony to the most heavenly illusion in the natural world.

* * * * *

Does anyone ever watch the partial eclipse after the end of totality? After third contact, all else is anti-climax. Bottles of champagne were uncorked. Passengers and crew celebrated. A photograph was taken of the 30 or so members of the party who had just lost their eclipse virginity.

The Swedish captain came on deck to join the celebrations, and to be included in the virgins' photograph. He was as thrilled as any of us, perhaps more so. After one of the expedition's leaders had formally thanked him and his crew for their successful manoeuvres around the Atlantic Ocean, the captain himself said a few words. He thanked Landon Noll and those who had spent the night refining their weather predictions, helping to plan strategy, and advising him, right to the last minutes, on the best way to orient the ship. He said that he had not anticipated the wonder, and when he set out on the cruise, he had been unable to understand why so many would travel so far for the experience. But now, having seen it for himself, he said he would have been happy to drive the ship for miles, as far as necessary, at no extra charge. He then thanks us for having made him do it! And when he got home, he would be planning a proper vacation to see one again.

His next chance would have been in March 2015, and in his own part of the world. The path crossed the northern extremes of the Atlantic Ocean, beyond the Arctic Circle, its only landfall being the Faroe Islands and Svalbard. Neither the Faroe Islands, which in my mind are associated with the shipping forecast and force ten storms in March, or Svalbard, at a temperature of minus something, seemed a high priority for a visit, although the sight of the northern lights at an eclipse would have been a remarkable experience. Also, an eclipse in the mist could resolve another conundrum: what are the colours of a rainbow during a total solar eclipse? The answer, in theory, is pink thanks to the spectral clues of coronal light, but this is an experiment that I have never seen reported. The chance of seeing one is probably infinitesimal. I decided to stay at home and watch an 85% partial eclipse. This would, in effect, be similar to my eclipse initiation, 61 years earlier in 1954.

When the 2015 eclipse happened, someone produced an interesting statistic about Europe's use of solar power. The obscuration of the sun throughout the continent on the morning of 20

March caused a loss of 35 terawatts of solar power—equivalent to the output of about ten conventional power stations. I can well believe it. In southern England, several hundred miles south of totality, the daytime turned to twilight and birds started to roost (Figure 9B).

I joined 200 fellow residents of south Oxfordshire and watched the eclipse in a park on the banks of the River Thames. The local astronomical society had set up telescopes, binoculars, and demonstrations for the children. An overcast day at the start of spring was cool even before the eclipse got underway, so there was no

Figure 9B. 'The crow and the crescent', an 80% partial eclipse in Abingdon-on-Thames, 2015. This eclipse was total in the vicinity of the Faroe Islands (photo: Frank Close). See Plate 12 for a colour version.

dramatic sense of additional cooling but, that apart, my memories of 1954, when my quest began, flooded back.

My grandsons, aged three and five, saw the eclipse in their school and playgroup in Surrey. 'The moon got in the way, grandpa!' they explained. Roll on 21 August 2017. We will breakfast in America, and then watch, somewhere along the track from Oregon in the north-west to South Carolina in the south-east, while the sun is eclipsed by the moon.

10

Back to the Future

At dawn on 21 August 2017, if you're in the vastness of the North Pacific Ocean about 1500 miles north-west of Hawaii, or 800 miles from the nearest land, Midway Atoll—39.75° N and 171.5° W to be precise—you could experience a truly bizarre natural phenomenon. On the eastern horizon, the sun and moon will appear in perfect alignment: a totally eclipsed sun will rise.

If any reader has witnessed this uncanny experience, I would love to hear of it. If the sun is totally eclipsed, what means sunrise? The event sounds like a natural oxymoron.

The rich orange colours, which normally herald the dawn, will not happen. In the moments immediately before sunrise, a galaxy of stars will shine in darkness deeper than that of the blackest night. Perhaps a vague shimmer of coronal silver will betray the point where the conjoined sun and moon are about to appear. The flash of a diamond ring will signal the day.

Every total eclipse on earth begins at someone's sunrise. To understand why, imagine the situation immediately before this singular moment. The moon's shadow is gradually swinging across space, like a negative version of a lighthouse beam. Suddenly, it makes its first contact with the surface of the earth, at which point the eclipse begins. The beam of darkness continues to swing across the globe, for 8200 miles, until it misses the planet and heads off into space once again. Its angle relative to the surface starts at grazing contact, grows to a maximum, and then dies back to nothing. Were you to be on the surface, under darkness, the line of sight to the eclipsed sun would be high up in the middle of the path, but along the horizon at its ends. Thus an eclipse begins at sunrise, somewhere in the west, and travels across the globe from west to east until sunset (Figure 10A). In the eclipse of August 2017,

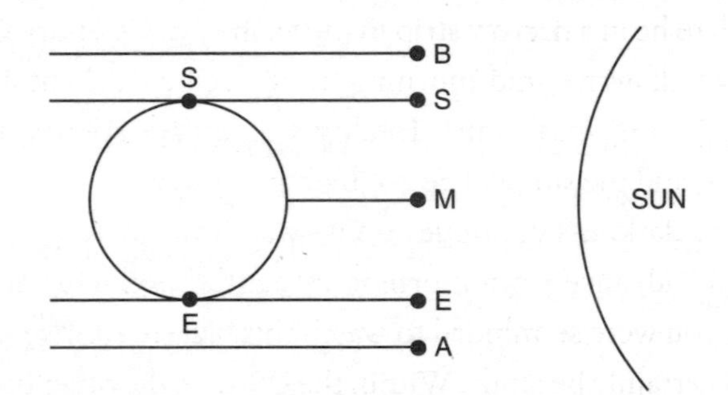

Figure 10A. Eclipse from dawn to dusk. The moon traverses from the top to the bottom of the diagram. Before the eclipse (B) its shadow misses the earth. The moon's shadow intercepts the earth first at S. A person on the earth who has just been spun from night time to the point S will see the eclipsed sun on the horizon: this is normal sunrise. The moon's shadow follows the moon's trajectory. An observer at M will see the eclipsed sun high in the midday sky. At E a person sees it at normal sunset. From this instant onwards, the moon's shadow misses the earth and goes into space.

the shadow will trail across the globe from sunrise in the Pacific, until it ends, at sunset in the Atlantic Ocean.

* * * * *

In the early morning of the eclipse's day, the moon's shadow will race across the ocean towards the western coast of the USA, 2500 miles away. By ship this journey might take a week; by plane, several hours. The moon's umbra travels at supersonic speed, and arrives on the coast of Oregon within 28 minutes.

Being further east, the landlubbers' day will have already lasted several hours. On the west coast of North America from Mexico up to Canada the eclipse will begin around breakfast time. Most will experience only a partial eclipse, however. For those fortunate enough to be in a narrow strip in the vicinity of Newport, Oregon, totality will arrive mid-morning: 10.16 Pacific Daylight Time, or 17.16 UT—'universal time'. Totality will last for approximately 2 minutes, and the sun will be 40° high in the sky.

As the darkness continues on its way from west to east, local time will advance from morning into afternoon. Out in the Pacific, if you were so minded to watch this bizarre sunrise, you will almost certainly be alone. Within the USA, on the other hand, you will probably be one in more than a hundred million who gather to witness the phenomenon.

* * * * *

There's plenty of choice for where to be. Between landfall at Fogarty Creek State Park in Oregon, on the west coast, to departure over the Cape Romain National Wildlife Refuge in South Carolina on the east, is exactly 2500 miles (Figure 10B). Together with its traverse over the Rocky Mountains, these locations hint at the intrinsic beauty of the scenery en route, eclipse or not.[1]

The width of the moon's shadow is about 63 miles in Oregon. If you are on the centre line at that location, totality will last for about 2 minutes. By the time it reaches South Carolina, the width is about 72 miles and the time extended to two and a half minutes. How has an increase in width by a factor of 1/7 elongated totality by 25%? This is due to the different orientation in the sky as we pass from morning to mid-afternoon, and the associated changing projection of the shadow onto the earth's surface.

The moon's cylindrical shadow hits the earth's surface at an angle, which forms an ellipse. My description of distances referred to the perpendicular to the track; the time of totality at any point is more a measure of the distance parallel to the path. The precise details of what follow will vary along the track as the projection, and hence the ellipse's eccentricity—oblateness—changes. Nonetheless, the qualitative conclusions remain the same.

If you are located near the centreline, more of the shadow will pass over you than if you are miles away, near the edge of the rim of darkness. You can be several miles off-centre, and still experience a long total eclipse. However, were you to approach the boundary, the time would fall off rapidly. To illustrate how, and to help you decide how much care you need when choosing your preferred location, let's imagine we were about 40 miles inland, in Oregon.

[1] For an interactive map of the path, which will be updated as the event approaches and more up to date information is known, try http://eclipse.gsfc.nasa.gov/SEgoogle/SEgoogle2001/SE2017Aug21Tgoogle.html.

Figure 10B. Eclipse of 21 August 2017. The path of totality across the United States is shown in grey. Outside this track, the sun will be partially eclipsed. (Map courtesy of Xavier Jubier, IAU Working Group on Solar Eclipses. See Plate 13 for a colour version.

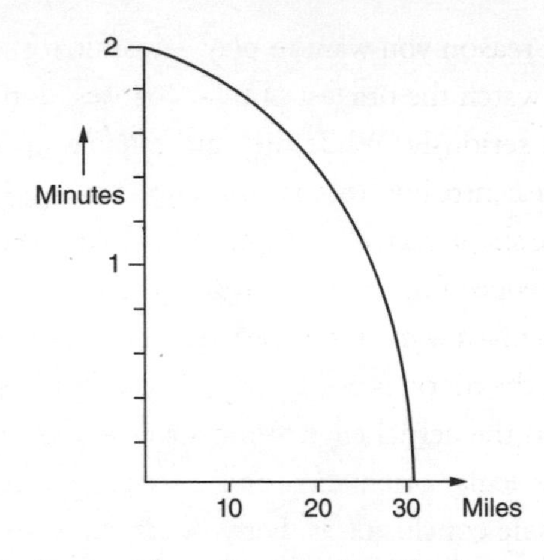

Figure 10C. Totality away from the centre line. The diagram illustrates how the duration of a total eclipse falls from 2 minutes at the centre line to zero at the boundary, some 31.3 miles away. There is a gradual decrease initially, which becomes more rapid after 20 miles.

I have chosen this location because on the centre line totality lasts for 2 minutes exactly, which makes for easy numerical comparisons. You have 31 miles of leeway before you are in danger of leaving the belt of totality (Figure 10C). As you leave the centre line, for the first ten miles you will hardly experience any change, as you will lose less than 1 second of totality per mile. Even at 20 miles out, which is two-thirds of the way to the edge, you will still have one and a half minutes of totality. The time reduces to 1 minute, half of the central amount, only if you venture 27 miles. From here out, however, the time drops rapidly: 35 seconds at 30 miles; 18 seconds at 31 miles. Stray a further 500 metres and you will be down to 10 seconds; another 100 metres and you will be out of range entirely.

If for some reason you want to play a cosmic form of Russian roulette, and watch the briefest of total eclipses, don't take these numbers too seriously. While they are reliable up to about 25 miles from the centre line, they are inaccurate as you approach the edge. First, the shape is an ellipse, not a circle. Also the precise trajectory of the edge of the moon's shadow will only be determined at an accuracy of a few metres in the weeks before the actual event. Furthermore, the moon is not a perfect circle, but has valleys and mountains. So the actual edge of the shadow will vary depending on the particular orientation of the moon as the eclipse progresses. The safe conclusion is that you can stray 20, perhaps 25 miles from the central line of the eclipse and still experience an extended total eclipse with all the emotion undimmed. Beyond that: caveat emptor!

Now that you have some idea of your lateral freedom, the main choice will be where to go along the 2500 miles of the track across the USA.

You could go off shore, near Lincoln City, Oregon, and—coastal fog permitting—see the sun vanish. One minute 58 seconds later, the diamond ring will herald the end of totality, and the moon's shadow will rush off eastwards to envelop the landmass of North America. From there on, perhaps a hundred million people will share the experience. Then, having arced across the Rocky Mountains, the Midwest, and Tennessee, the path of totality will leave the continental USA on the Atlantic coast of South Carolina an hour and a half later at 14.46 EDT. It will then cross the Atlantic, without further landfall, until it ends at sunset in the tropics, about 300 miles south of the Cape Verde islands, and as at sunrise, its duration shrunk to a second.

On the west coast of the USA, the eclipsed mid-morning sun will be some 40° in altitude. It will get as high as 64° in Tennessee and still be 61° above the horizon as it heads off into the Atlantic afternoon. On the shadow's centre line, totality will last 1 minute and 54 seconds on Oregon's Pacific coast, rise to a maximum duration of 2 minutes 40 seconds near Hopkinsville, Kentucky, before falling off slightly to 2 minutes 34 seconds as it exits the east coast in South Carolina, between Charleston and Georgetown.

If duration of totality is your metier, be near to the centre line and choose anywhere east of Kentucky. However, good weather is key, so bear in mind the golden rule: 'Better shorter in clear sky than longer under cloud.' As the eclipse is in August, the omens for fair weather are good. The summer thunderstorm season will be ending and the dry autumn on its way. On balance the west and Midwest have the best prospects, but as for conditions on the actual day, who knows: my experience in August 1999 testifies to that. However, modern weather forecasting is good, and once you have chosen your nominal base, you may be able to revise your preference in the days or even weeks ahead. The USA has an excellent network of roads, on which you can make last minute detours to chase a hole in the clouds, but bear in mind the 1999 experience of those in mainland Europe, trapped in traffic jams on the autobahn, or in Libya 2006, where the only road across the Sahara became choked with cars.

The USA has more roads than the Sahara—although roads can be sparse in some areas along the eclipse path, such as the Rocky Mountains—but it has vastly more automobiles. To add to the problem, there are several major conurbations within 50 miles of totality. It is almost certain that more people will attempt to see this eclipse than any other, not least its European forebear of 1999.

So get a map of the USA to hand, and check the eclipse path relative to the continental system of highways and byways. For the highways will bring in the hordes from the surroundings, but last minute tactics may depend on finding rat runs, away from the mainstream. For that you will need local knowledge and all I can offer here is to recommend that, once you have decided on your base-camp, you reconnoitre the area and plan your strategy.

I remember the build-up to Cornwall 1999, where the authorities worried lest a million people invaded Cornwall via the couple of major roads to the peninsula. What might a similar analysis suggest for 2017 in the USA?

* * * * *

The eclipse crosses from west coast to east, and thus traverses every major north-south interstate highway. It would be wise to use these in advance of the eclipse to reach your target, as there will be many who decide at the last moment to make a trip of 50 miles on the interstate, and even larger numbers might be ready to travel more than 200 miles. After all, how far would you be prepared to drive to see a once in a lifetime natural phenomenon within a day's journey of your home? The east coast of the USA is expert at reassigning directional flows on freeways at times of crisis, such as when hurricanes strike. 'Hurricane evacuation route' is a common signage in the Deep South, and up the coast of Virginia. 'Eclipse access', and subsequent evacuation, could be merited at several locations across the continent.

On the west coast, the path crosses I-5 in Oregon between Eugene and Portland. Up to a million people from those conurbations have less than 50 miles to reach totality. Where 50 miles can be a torture to drive in England, it is a commute in parts of the USA. One or two hundred miles is nothing more than a few hours' drive, with cheap gasoline. Suddenly totality is within the range of Seattle, normally 3 hours' drive along I-5. However, as in all the following, take account that you will not be alone in your chosen route to reach the eclipse. In some cases, there will probably be major traffic jams. The following also give some idea why estimates that the eclipse will have 100 million watchers are not an exaggeration.

From the south on I-5 the Bay Area is 450 miles. This could generate much traffic northbound overnight or pre-dawn. Coastal motels can expect good business. If that is your intention, book early.

The track crosses Oregon, where the High Desert east of the Cascade Mountains promises good prospects of clear skies. Then on it goes, through Idaho, north of Boise, and across the Rocky Mountains and Grand Teton National Park. The north–south interstate artery of I-15 gives access from Salt Lake City, 170 miles, and I-25 brings Boulder and Denver, Colorado, within reach at150 miles.

The shadow crosses 450 miles of Nebraska in just 17 minutes. Lincoln is just within the track, and can expect 1 minute 45 seconds of totality. Several conurbations are within a couple of hours drive, such as Omaha, less than 50 miles away on I-80, and Des Moines, 110 miles to the north on I-35. Kansas City straddles the southern edge of the totality band, so expect traffic heading north on I-29 and I-35, or east on I-70 for totality at 18.08 UT, 13.08 Central Daylight Saving Time.

Twenty-seven seconds later, St Louis will be on the northern boundary. Over two million in its metropolitan area can access over 2 minutes in the moon shadow with a trip southwards of a few miles. I-55 might not be the best choice, however, as that artery is likely to be clogged.

Here it is a matter of how far would-be eclipse chasers are prepared to travel. Springfield is 90 miles away on I-55, while the conurbation of Chicago at 260 miles is well within range. I-55 and I-57, via Champaign, are both likely to be dense with traffic. Whereas 5 hours would be sufficient on a normal day, bear in mind the experiences of those on the German autobahns in 1999 or in the Sahara Desert in 2006.

The path then passes through the south-west corner of Kentucky—Louisville is just 94 miles away on I-65—and into Tennessee, where Nashville gets nearly 2 minutes, and Memphis is 140 miles away on I-55. Indianapolis at 170 miles and Cincinnati at 180 miles add to the numbers.

As totality approaches the east coast, at about 2.30 in the afternoon, I-85 cuts across the track midway between Atlanta and Charlotte, a mere 70 miles to the south and north, respectively. Well over five million people will be within 50 miles of totality at that location alone. To add to the potential for stagnation on the roads, most of Florida is within a day's journey; Washington DC is less than 400 miles away, and northern Virginia 300.

A large population on the southern half of the east coast lives within a day's trip. Weather prospects are likely to be better in the western states, whereas the east has a tendency to cloud cover at that time of the year. If so, there could be much last minute movement as people seek better viewing conditions.

As I already hinted from the experience of 1999 in Germany, the interstate arteries are both opportunity and threat. If you are planning to travel that day, and have no interest in the eclipse, avoid these areas in the morning, at least. If you are a local, and need to reach a good site, use local knowledge and avoid the interstates. If you live elsewhere in the continental USA, you could be one of over 100 million for whom the path would normally be within a day trip by car. Thus if you plan to make the trip, be assured that you will not be alone. Leave early!

As for a back-up plan, in case you get stuck, the car breaks down, or it rains: this won't be your last chance, at least if you're in the USA. If you miss the eclipse of 2017, clear your diary for 8 April 2024.

* * * * *

As I said earlier, the eclipse of 2017 is in the same saros as the European eclipse of 1999, where my quest began. On 29 March 2006, I went to Libya for an eclipse, which came from the south-west, across the Sahara Desert, and then headed north-east across the Mediterranean and Turkey, where it crossed the track of the 1999 eclipse. Add 18 years and ten days for a saros, and we find: 8 April 2024. And as the 1999 eclipse trail is displaced some 120° to the west, so is that of the 2006 track in 2024. Totality in April 2024 will enter the USA from Mexico and head north and east, through Texas, up to Illinois and along the border of Canada in the far north-eastern corner of the USA.

Although at any random spot on earth, the average wait between two total eclipses is 400 years, the odds include winners and losers. The paths of the 1999 and 2006 eclipses crossed in Turkey, a coincidence that will also be translated 120° to the west for 2017 and 2024.

Citizens near the join of Illinois, Missouri, and Kentucky can stay home on both occasions. You are one of the lucky few for whom two total eclipses will occur in a mere seven years. The centre lines will cross in southern Illinois near Carbondale. The two million residents of St Louis will be on the edge of the path in both cases. Join the throng; it would be such a waste to live just outside two total eclipses and experience neither.[2]

[2] To choose your ideal spot for 2017, and assess the weather prospects along the track through Oregon, Idaho, Wyoming, Nebraska, Missouri, and the Carolinas, start with http://www.eclipse2017.org/2017/path_through_the_US.htm.

Epilogue: Everything under the Sun is in Tune

From my virginal eclipse of 11 August 1999 to that in the USA on 21 August 2017 is the classic 18 years and ten days of one saros. This is an appropriate place for me to stop and to pass on the baton.

My personal journey has been the legacy of Mr Laxton—Cyril as I later came to know him. He opened my eyes to natural wonders and helped frame the course of my life. Back in the halcyon days of 1954, his summary of totality was that the experience would be unforgettable, and that when I had seen one for myself, I 'would know what he meant'. Six decades later I can confirm: yes, I do. I hope that, having read this, you too now have some sense of the wonder, and will be inspired to 'see one for yourself'.

Cyril Laxton had seen totality on 29 June 1927. He had shared the partial eclipse of 1954 with his students, and given me the inspiration to see the total eclipse on 11 August 1999. As you might have noticed already, between 27/6/27 and 11/8/99 there are 72 years and 43 days: four saros.

For me in 1999, of course, the irony was that the main actors were completely obscured by thick cloud. Only as totality ended did the clouds begin to thin, but too late. One minute after that blackout ended, the moon's shadow had moved on about 100 miles and made its last contact with the British mainland.

Some years later, I learned what happened next.

That final landfall of the moon's shadow enveloped the town of Torquay, on the English Riviera—Devon's south coast. Its residents managed to experience about 1 minute of a total solar eclipse, and under relatively clear skies. Among them, in retirement aged 91, was Mr Cyril Laxton.[1]

[1] Cyril Laxton died a few days short of his 99th birthday in 2007.

INDEX